CAPTURING

THE UNIVERSE

The most spectacular astrophotography
from across the Cosmos

RHODRI EVANS

ANDRE
DEUTSCH

CONTENTS

FOREWORD by Steven Young, Editor *Astronomy Now*

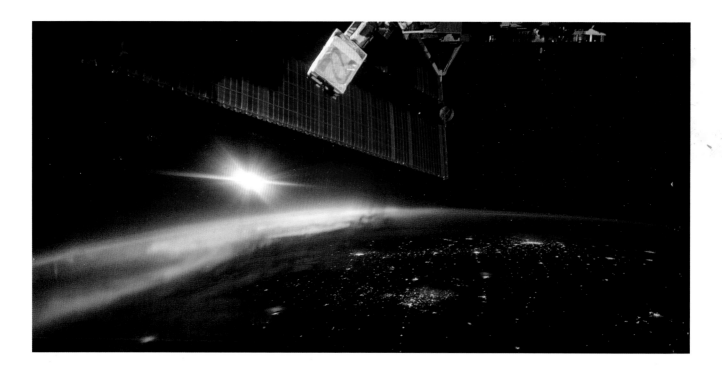

Since the very first humans gazed up at the night sky, we have been fascinated by the beauty of the heavens above. Early civilizations struggled to make sense of the sweeping tapestry of stars. The patterns painted by these brilliant pinpoints of light were named. Myths and legends were based on what they saw.

The early stargazers could never have anticipated that their descendants would one day shape glass and mirrors and look beyond the stars visible to them, and deep into the cosmos. Or that they would leave footprints on the Moon and build machines to explore the wandering stars we now know to be the planets circling our own Sun, which itself is just one example of the seemingly infinite stars in the sky.

Over the centuries theories came and went about the nature of the stars and the planets. Understanding the motions of these heavenly bodies became important for navigating our planet by sea but our curiosity about what lies out there has driven our quest to see the cosmos in ever greater detail.

The arrival of photography in the 1800s opened a new window on the Universe. Long exposures on photographic film were capable of capturing more photons than the human eye could ever see, allowing us to observe nebulae and distant galaxies like never before. The dawn of the digital age has opened our eyes to the Universe in unprecedented detail. Imaging chips, far more sensitive than film, have flown aboard space probes to the far reaches of our Solar System. They orbit high above our planet on observatories like the Hubble Space Telescope, enjoying views undistorted by the atmosphere.

In this book, Dr Rhodri Evans presents the best astronomical images gathered by robotic space explorers and ground-based telescopes and explains the science behind the beauty. For those who appreciate the treasures of the cosmos, there has never been a better time to be alive.

Above: NASA astronaut Scott Kelly captured this photograph of the green lights of the aurora from the International Space Station on 7 October 2015.

TELESCOPES AND SPACECRAFT

Telescopes

Anglo-Australian Telescope (AAT) Completed in 1974, the 3.9-metre AAT is located at Siding Spring Observatory in New South Wales, Australia. Built as a collaboration between the United Kingdom and Australia, it has been entirely operated by Australia since 2010. Until the Blanco telescope was completed in 1976, the AAT was the largest telescope in the Southern Hemisphere. Siding Springs Observatory is at a relatively low altitude of 1,100 metres, but boasts one of the darkest skies in the world due to its remote location. The AAT has been ranked as the fifth most scientifically productive telescope based on the research output from data obtained using the telescope.

Arecibo Observatory (Arecibo) This is the largest radio telescope in the world, built into the crater of an extinct volcano in Arecibo, Puerto Rico. Arecibo's single fixed dish is 305 metres in diameter; suspended above the dish are the instruments that detect and measure the radio waves. The instrument housing can be moved to allow different objects to be observed as they transit the local meridian. Arecibo was completed in 1963, and operates from frequencies of 300 Mega Hertz to 10 Giga Hertz (wavelengths from three centimetres to one metre).

Atacama Large Millimetre Array (ALMA) At an altitude of 5,000 metres on the Chajnantor plateau of the Atacama Desert in northern Chile, the Atacama Large Millimetre Array takes advantage of the exceptionally dry air at such altitudes to make observations at millimetre wavelengths, radiation which does not reach sea level. ALMA is operated by a group of countries that currently includes: the United States of America, the countries of the European Southern Observatory (Belgium, Germany, France, the Netherlands, Sweden, Denmark, Switzerland, Italy, Portugal, the United Kingdom, Finland, Spain, Czech Republic, Austria and Poland), Canada, Chile, and Japan. ALMA began operation in 2011 and, when complete, it will consist of 66 12-metre and 7-metre millimetre-wave dishes.

Blanco Telescope (Blanco) This 4-metre telescope is located at the Cerro Tololo Inter-American Observatory (CTIO) in Chile, at an altitude of 2,207 metres. Named after Puerto Rican astronomer Victor Manuel Blanco, it went into operation in 1976. From its completion until 1998, Blanco was the largest telescope in the Southern Hemisphere, when ESO's VLT first saw light. The CTIO is operated by the National Optical Astronomy Observatory (NOAO), the United States' national observatory for ground-based astronomy. The NOAO operates CTIO, Kitt Peak National Observatory in Arizona, and the two 8-metre Gemini telescopes located on Mauna Kea in Hawaii and Cerro Pachón in Chile.

Canada-France-Hawaii Telescope (CFHT) The Canada-France-Hawaii telescope is a 3.6-metre visible and infrared light telescope jointly operated by the French national science research centre (CNRS), the National Research Council of Canada and the University of Hawaii. It also has contributions from China, Taiwan, Brazil and South Korea. The CFHT went into operation in 1979 and is located on Mauna Kea in Hawaii, at an altitude of 4,204 metres above sea level.

Chandra X-ray Observatory (Chandra) One of NASA's Great Observatories, the Chandra X-ray Observatory was launched aboard the Space Shuttle Columbia in July 1999. Named after Indian-born theoretical astrophysicist Subrahmanyan Chandrasekhar, it is a 1.2-metre telescope optimized for X-rays between wavelengths of 12.4 nanometres and 124 nanometres. Chandra is in a highly elliptical orbit around the Earth, varying between an altitude of over 130,000 kilometres to a much lower 14,000 kilometres. Although its planned lifetime was only five years, Chandra is still in operation.

Herschel Space Observatory (Herschel) With a 3.5-metre mirror, the Herschel Space Observatory was an infrared telescope and remains the largest telescope ever put into space. Developed by the European Space Agency, Herschel was launched in May 2009 and put into a Lagrangian-2 (see Glossary) orbit, 1.5 million kilometres from Earth. Herschel operated in the infrared between wavelengths of 55 microns and 672 microns. Named after William Herschel, who discovered the infrared in 1800, the instruments ran out of liquid helium coolant in April 2013, whereupon Herschel ceased operation.

Hubble Space Telescope (HST) Put into Earth-orbit by the Space Shuttle Atlantis in April 1990, the Hubble Space Telescope is possibly the most famous telescope ever. Named after Edwin Hubble, it has a 2.4-metre diameter mirror and operates in the visible, ultraviolet and near-infrared parts of the spectrum. HST orbits the Earth at an altitude of about 540 kilometres to get above the blurring effects of the Earth's atmosphere. Soon after it started taking its first images it was realized that the primary mirror had not been made to the correct shape. In December 1993 astronauts aboard the Space Shuttle Endeavour installed an optics package to correct this problem. Its original suite of five scientific instruments has been replaced on a regular basis, to keep its scientific capabilities at the cutting edge of technology. HST is one of NASA's Great Observatories, along with the Compton Gamma Ray Observatory, the Chandra X-ray Observatory and the Spitzer Space Telescope.

James Clerk Maxwell Telescope (JCMT) The JCMT is named after Scottish theoretical physicist James Clerk Maxwell, who derived the equations for electromagnetism and showed that light is just one part of a much broader phenomenon, what we now call the electromagnetic spectrum. The JCMT is located on the summit of Mauna Kea in Hawaii, at an altitude of 4,092 metres, and has a 15-metre diameter dish designed to work at sub-millimetre and millimetre wavelengths. It takes advantage of atmospheric windows at 450 and 850 microns which open up when the air above the telescope is particularly dry. The JCMT went into operation in 1987 and was funded until February 2015 by the United Kingdom, Canada and the Netherlands. Since March 2015 its operation has been taken over by the East Asian Observatory, jointly funded by China, South Korea, Japan and Taiwan.

Keck Telescopes (Keck) The two Keck Telescopes (Keck I and Keck II) are located on the summit of Mauna Kea in Hawaii, at an altitude of 4,145 metres. Jointly operated by the University of California and the California Institute of Technology (Caltech), the Keck telescopes observe in the visible and near-infrared parts of the spectrum. Their ten-metre diameter mirrors are each formed from 36 hexagonal segments; Keck I went into operation in March 1993 and Keck II in January 1996. They currently are the largest steerable visible/near-infrared telescopes on Earth, but their pioneering segmented-mirror design will see them overtaken in the 2020s by a new generation of 30-metre telescopes currently being planned.

Max Planck Gesellschaft Telescope (MPG) This is a 2.2-metre telescope located at La Silla Observatory in Chile and operated by the European Southern Observatory. Construction of the MPG was funded by the Max Planck Society (*Max Planck Gesellschaft* in German), and completed in 1984. It was on loan to ESO from the Max Planck Institute for Astronomy (MPIA); in October 2013 the MPG was returned to the MPIA. Time on the telescope is shared between the MPIA and the Max Planck Institute for Extraterrestrial Physics (MPE); however, ESO remains responsible for its operation and maintenance. Currently MPG hosts three instruments: a 67-million pixel Wide Field Imager with a field of view as large as the full Moon, a camera for detecting gamma-ray bursts, and a high-resolution spectrograph used for making detailed studies of stars.

Mount Wilson 100-inch Telescope (100-inch) Next to the Hubble Space Telescope, the 100-inch Telescope at Mount Wilson Observatory (MWO) is one of the world's most famous telescopes. Completed in 1917, the 100-inch was, until 1948, the largest visible light telescope in the world. The 100-inch was the telescope used by Edwin Hubble in the 1920s to show that our Galaxy was not the entire Universe, and later that the Universe was expanding. Located on Mount Wilson, which is in the San Gabriel Mountains to the east of Los Angeles, MWO is at an elevation of 1,740 metres. Until the 1940s it was the premier observatory in the world, but its usefulness for observing deep-sky objects became increasingly hampered by the light pollution of Los Angeles. However, due to the exceptionally stable air above MWO, it is still used, primarily to develop adaptive optics systems.

Palomar 48-inch Schmidt Telescope (48-inch) Completed in 1948 to complement the new 200-inch Mount Palomar Telescope, the 48-inch uses a Schmidt corrector plate to provide an extremely wide field of view ideal for sky surveys. Starting in 1949, the 48-inch conducted several photographic surveys of the entire sky visible from Palomar Observatory, located to the east of San Diego in Southern California. These surveys are known as the Palomar Observatory Sky Survey (POSS), and comprise nearly 2,000 photographic plates. The survey was completed in 1958, using both blue-sensitive and red-sensitive photographic plates. In the 1980s and 1990s a second-generation survey was conducted, POSS-II, using more sensitive photographic films at blue, red and near-infrared wavelengths. The telescope has since been converted to use an array of charge-coupled devices (CCDs). A mosaic of 12 CCDs currently give the 48-inch the largest field of view of any telescope using CCDs.

Spitzer Space Telescope (Spitzer) Launched in August 2003, the Spitzer Space Telescope (Spitzer) is an infrared telescope with a 0.85-metre mirror. Operating between wavelengths of 3.6 microns and 100 microns, the mirror and instruments were initially cooled using liquid helium to a temperature of 5.5 Kelvin to make the telescope more sensitive. The liquid helium supply was exhausted in 2009, but Spitzer continues to operate using its shortest wavelength camera. Spitzer was not placed in orbit around the Earth, but rather in an Earth-trailing orbit where it orbits the Sun, slowly drifting away from the Earth. Spitzer was named after astronomer Lyman Spitzer, who promoted the idea of telescopes in space in the 1940s, before the advent of the space age. It was the last of NASA's Great Observatories to be launched.

Subaru Telescope (Subaru) This is an 8.2-metre visible and infrared light telescope operated by the National Astronomical Observatory of Japan and located at an altitude of 4,139 metres on the summit of Mauna Kea in Hawaii. Subaru went into operation in 1998, and is named after the Japanese word for the Pleiades open star cluster.

Very Large Array (VLA) Located in New Mexico, the Very Large Array comprises 27 25-metre radio dishes that form a Y-shaped array. The telescopes operate between frequencies of 74 Mega Hertz and 50 Giga Hertz (wavelengths from 0.7 centimetres to 400 centimetres). The VLA is possibly most famous for featuring in the movie *Contact*, starring Jodie Foster.

Very Large Telescope (VLT) The Very Large Telescope comprises four 8.2-metre telescopes operated by the European Southern Observatory. The VLT is located on Cerro Paranal in the Atacama Desert of northern Chile, at an altitude of 2,635 metres above sea level. The telescopes operate at visible and near-infrared wavelengths, and went into operation in 1998. The VLT can either be operated as four separate telescopes, or their light can be combined to produce much higher resolution images.

Visible and Infrared Survey Telescope for Astronomy (VISTA) Built to provide wide-field images, the 4.1-metre VISTA telescope is located at Paranal Observatory in Chile, at the same location as the VLT. Unusually for modern-day telescopes, it is only used with one instrument, the Vista InfraRed CAMera (VIRCAM), which contains 16 detector-arrays sensitive to near-infrared light, from about 0.85 microns to about 2.15 microns. The 16 detector-arrays have a total of 67 million pixels, and together cover an area of 0.6 square degrees on the sky (for comparison, the full Moon covers an area of 0.25 square degrees). Like the VLT, VISTA is operated by the European Southern Observatory and went into operation in 2009.

Wide-field Infrared Survey Explorer (WISE) Launched in December 2009, WISE was an infrared space telescope designed to survey the entire sky at four infrared wavelengths, 3.4, 4.6, 12 and 22 microns. With a 0.4-metre aperture, it was placed in a low Earth-orbit with an altitude of just under 500 kilometres. The sky survey was completed in six months, and after a further three months the coolant used to cool the instruments to 17 Kelvin ran out. WISE was put into hibernation in February 2011. However, in August 2013 NASA decided to re-activate WISE to look for asteroids, with the detectors working at ambient temperatures with a much lower sensitivity.

Spacecraft

Cassini Launched in October 1997 to orbit and study Saturn, her ring system and her moons, the Cassini spacecraft arrived in July 2004. Since its arrival, Cassini has studied the Saturnian system in unprecedented detail, including numerous fly-bys of several of its moons such as Enceladus, Hyperion and Titan. Cassini is still in operation, but the mission is currently expected to come to an end in 2017. The Cassini spacecraft was named after Italian astronomer Giovanni Cassini, who in 1675 discovered the gap in Saturn's rings, which is now known as the Cassini Division.

Curiosity Rover With a length of 2.9 metres, a width of 2.7 metres and a height of 2.2 metres, the Curiosity Rover is the largest rover to be put on the Martian surface. Curiosity landed on 6 August 2012 in the Gale Crater on Aeolis Palus; its landing coordinates were 4.6 degrees South and 137.4 degrees East.

As on Earth, the choice of zero latitude is determined by the planet's equator, but the choice of the Martian prime meridian is as arbitrary as the choice of prime meridian (Greenwich) on Earth. German astronomers Wilhelm Beer and Johann Heinrich Mädler selected a small circular feature on the Martian surface when they produced the first systematic charts of Martian features in 1830–32. This was adopted in 1877 as the prime meridian for Mars, later refined to the Airy-0 crater in the Sinus Meridiani in the 1970s.

With a mass of 900 kilogrammes, Curiosity is equipped with a suite of scientific instruments including cameras, spectrometers, a scoop to take soil samples and analyse them in a chemical laboratory on-board the lander, a

laser to vaporise rock and even a drill which can drill into rock up to a depth of five centimetres. Curiosity had a target mission lifetime of 668 sols, but it had passed this target by mid-2014. Unlike its predecessors Spirit and Opportunity, which were powered by solar panels, Curiosity is powered by a radioactive isotope. In December 2012 its mission was extended indefinitely, and during its primary two-year mission it travelled about 20 kilometres.

Galileo Sent to study Jupiter and its four largest moons, the Galileo spacecraft was launched in October 1989. It arrived at Jupiter in December 1995, and spent nearly eight years studying the Solar System's largest planet and its four largest moons, Io, Europa, Ganymede and Callisto. Galileo ended its mission in September 2003; in order not to accidentally crash into any of the moons and disturb their pristine environment, it was sent into the Jovian atmosphere and burned up there. The spacecraft was named after Galileo Galilei, who discovered the four largest moons in January 1610.

Huygens Named after Dutch astronomer Christian Huygens who discovered Saturn's largest moon Titan in 1655, the Huygens spacecraft was transported to Saturn by the larger Cassini spacecraft. Huygens separated from Cassini on 25 December 2005, and on 14 January 2006 it entered Titan's atmosphere, successfully landing on its surface. Huygens is the first lander to land on the surface of a body in the outer Solar System, beyond the asteroid belt. During its descent, Huygens measured Titan's atmosphere, including its composition, temperature and pressure as a function of altitude, as well as taking photographs during its descent and after landing. At the surface, Huygens measured the physical properties of the surface at the landing site, including whether it was solid or liquid. Huygens continued to send data for 90 minutes after it landed, exceeding its design specifications.

MESSENGER This is an acronym for the MErcury Surface, Space ENvironment, GEochemistry, and Ranging spacecraft, launched from Earth in August 2004. MESSENGER was only the second spacecraft to be sent to Mercury; previously Mariner 10 was launched in November 1973 to perform close fly-bys of the planet. Mariner 10 performed three fly-bys of Mercury, in March 1974, September 1974 and March 1975.

MESSENGER first flew past Mercury in January 2008, then again in October 2008 and September 2009. After this, MESSENGER was manoeuvred into an elliptical orbit about Mercury, becoming the first spacecraft to orbit the planet. It orbited Mercury from March 2011 until April 2015, at which point MESSENGER had insufficient fuel to maintain its orbit. It was crashed into the surface of Mercury on 30 April 2015.

New Horizons Launched in January 2006 to visit the dwarf planet Pluto, New Horizons arrived at its target in July 2015. It passed Jupiter in February 2007, performing a gravity assist to help it reach Pluto. After its encounter with Jupiter, New Horizons was put into hibernation to conserve power; it came out of hibernation in December 2014. New Horizons began its approach phase to Pluto in January 2015, and on 14 July 2015 it flew 12,500 kilometres above the surface of the planet. It has now been manoeuvred to perform a fly-by of the Kuiper belt object 2014 MU, which should happen in January 2019.

Pioneer 10 and 11 These were the first spacecraft to be sent to explore the outer Solar System beyond Mars. Pioneer 10 was launched in March 1972, Pioneer 11 in April 1973. Pioneer 10 flew past Jupiter in November 1973, but did not encounter any of the other outer planets. Communication with Pioneer 10 was lost in January 2003 due to a loss of power to its radio transmitter. At this time, Pioneer 10 was about 12 billion kilometres (80 Astronomical Units) from Earth. In January 2016 it was calculated to be about 16.9 billion kilometres (114 Astronomical Units) from Earth, and is heading in the direction of the constellation Taurus.

Pioneer 11 flew past Jupiter in November and December 1974, and past Saturn in September 1979. Contact with Pioneer 11 was lost in September 1995 but, based on its speed and trajectory, it was calculated in 2015 to be about 13.5 billion kilometres (91 Astronomical Units) from Earth. It is heading towards the constellation Scutum.

Spirit and Opportunity Rovers The Spirit and Opportunity Rovers landed on the surface of Mars on 4 January 2004 and 25 January 2004 respectively. They became the second and third rovers to explore the surface of Mars, building on the success of the Sojourner Rover, which landed on Mars in July 1997. Sojourner was active for only 83 sols, and during this time it travelled just over 100 metres.

Spirit landed near the crater Gusev, a giant impact crater, which may at one time have contained liquid water. Its landing coordinates were 14.6 degrees South and 175.5 degrees East. Spirit was active from January 2004 until late 2009, when it became stuck in soft soil. Its last communication with Earth was 22 March 2010, but during these six years of operation Spirit travelled 7.73 kilometres.

Opportunity landed in Meridiani Planum with coordinates 1.9 degrees South and 354.5 degrees East, on the other side of Mars to Spirit's landing site. By chance, Spirit landed in an impact crater, and since its landing in January 2004 it has travelled over 40 kilometres. As of 2016, Opportunity continues to operate, exceeding its planned mission lifetime of 90 sols by over 12 years!

Viking 1 and 2 These were the first space probes to successfully land on the surface of Mars. Launched in August 1975, Viking 1 landed in western Chryse Planitia on 20 July 1976; the coordinates of the landing site were 22.5 degrees North, 50.0 degrees West. Viking 2 was launched in September 1975, and it landed at Utopia Planitia on 3 September 1976; the coordinates of its landing site were 50.0 degrees North, 225.7 degrees West (on the other side of Mars to Viking 1). The Viking landers were transported to Mars by their respective orbiters, which remained in orbit while the landers performed their experiments on the surface. The orbiters took images of potential landing sites prior to the landers being separated, and once the landers were on the surface the orbiters made measurements of the Martian atmosphere and relayed the data from the landers back to Earth.

The Viking landers performed a number of experiments, including measurements of the Martian atmosphere during their descent, and studies of the Martian surface including its geology and composition. Each Viking lander had a scoop that took soil samples and baked the soil in on-board ovens to look for signs of life. The Viking 1 lander operated for 2,307 Earth-days (or 2,245 sols), Viking 2 operated for 1,316 Earth-days (1,281 sols).

Voyager 1 and 2 These spacecraft were launched in 1977 to take advantage of an alignment of the outer planets (Jupiter, Saturn, Uranus and Neptune), which would enable a single spacecraft to visit all four planets. Voyager 1 was launched in September 1977, Voyager 2 two weeks earlier in August 1977.

By following a shorter trajectory, Voyager 1 arrived at Jupiter before Voyager 2, in January 1979. Voyager 1 then went on to Saturn, arriving in November 1980, and then flew past Titan (Saturn's largest moon). Its trajectory past Titan took Voyager 1 out of the plane of the ecliptic, and so it could not visit Uranus or Neptune. Voyager 1 continues to travel out of the Solar System, and it is currently at a distance of about 20 billion kilometres (134 Astronomical Units), which makes it the most distant man-made probe from Earth. Voyager 1 uses a radioactive source to provide electrical power, and is expected to continue operating until about 2025.

Unlike Voyager 1, Voyager 2 was sent on a trajectory that would take it past all four outer planets. It arrived at Jupiter in July 1979, Saturn in August 1981, Uranus in January 1986 and Neptune in August 1989. It is currently about 16.5 billion kilometres (110 Astronomical Units) from Earth, and like Voyager 1 it is expected to continue transmitting radio signals until at least 2025.

NOTE ON UNITS

Throughout this book we have used the metric system, which forms the basis of the SI (Système International) units used in science. In the SI system, metres are the units of length, kilogrammes are the units of mass and seconds are the units of time. We have also adopted the definition of billion used in science, which is a thousand million. (Historically in Britain a billion was a million million, but this usage is no longer commonly used.)

We have also adopted the common prefixes used in science to denote thousands, millions, thousandths etc. These are

· Kilo – thousand (as in kilometre, kilogramme)
· Mega – million
· Giga – billion (one thousand million)
· milli – thousandth (as in millimetre)
· micro – millionth
· nano – billionth

The word "micron" is often used as a contraction of micrometre, and denotes a millionth of a metre.

The exceptions to our use of SI units is for distances. Because distances are so huge in astronomy, using metres or even billions of metres becomes cumbersome. On the scale of the Solar System, it is common to express distances in Astronomical Units (AUs). One AU is defined as the average distance between the Earth and the Sun, an AU is 149.6 million kilometres, or 149.6 billion metres.

For distances beyond our Solar System, even AUs become too small. The most commonly used unit of length for distances beyond our Solar System is the light-year. This is defined as the distance that light travels in a vacuum in one year, and is equal to approximately 9 trillion kilometres (about 9 million million metres). Astronomers actually prefer to use a different unit, the parsec. The parsec is defined in the glossary, and is equal to approximately 3.3 light-years.

INTRODUCTION

PALE BLUE DOT

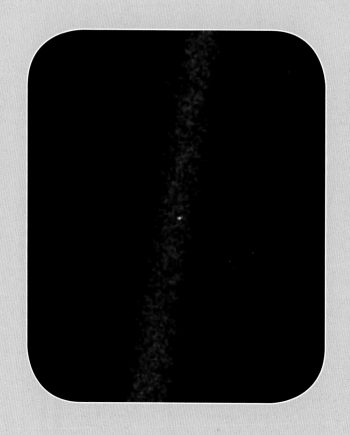

This picture shows the most distant image of planet Earth ever taken. It was obtained by the Voyager 1 space probe on 14 February 1990; at the time Voyager 1 was about 6 billion kilometres from Earth. The tiny dot which lies in a beam of sunlight (created by internal reflections in the camera) is our home. Everything we have ever experienced; all of history, all of the people you have ever known or loved or heard of, they have all lived on this tiny dot. The Earth covers less than a pixel of the camera's array, any further away and we would not have been able to see it at all. This is a truly humbling picture that shows our fragility and insignificance in the cosmos.

The image was the idea of Carl Sagan, the planetary scientist renowned since the 1970s for his landmark TV series *Cosmos*. Voyager 1 continues to travel further away from us, reaching the very edges of the Solar System. Maybe, one day in the future, human beings will leave our Solar System and travel to the nearest stars. The furthest human beings have travelled to date is to our nearest neighbour, the Moon, which is so close to the Earth in this image that it occupies the same fraction of a pixel as our home planet.

Yet, for thousands of years we, the occupants of this pale blue dot, have gazed out into the night-time sky and wondered at what lies beyond our small planet. Slowly, over many centuries, this wonder has been added to by an increasing understanding. Initially through the development of the telescope in the early 1600s, and later through spectroscopy in the mid-1800s and other developments, we now understand the true scale of the Universe and our place in it. Our planet is just one of eight in orbit around an average star, a star which is just one of hundreds of billions in our Milky Way Galaxy. Our Galaxy is just one of hundreds of billions of galaxies visible in our Universe, which we now know started 13.8 billion years ago. There seems to be nothing remarkable about our Earth at all, except it is special because it is our home. And, as of now, it is the only planet we know of which harbours life and us, a species who through our insatiable curiosity has found ways to explore the Universe in which we live.

This book presents some of the most beautiful visual highlights of what we have learnt about the Universe. Many of the images are obtained at wavelengths outside the visible part of the spectrum. Images in X-rays, radio waves, infrared light and microwaves can increase our understanding of an object. We start off with the Solar System and then examine our home Galaxy, the Milky Way, before looking at the neighbourhood of our Galaxy and its companions in the Local Group of galaxies. Beyond our Local Group lies a vast array of different types of galaxies. And, as we look even further away we see back to how our Universe was when it was young, observations which have taught us so much about the origins of our Universe and our very own origins.

Following pages: The Elephant's Trunk nebula is a cloud of gas and dust about 2,400 light-years away from Earth and more than 20 light-years long. It is part of a larger region of ionized gas in the constellation Cepheus, illuminated by a nearby massive star, which is off to the left of the area shown in this image. New stars are being born inside the Trunk. This image was taken by the Isaac Newton Telescope on La Palma in the Canary Islands.

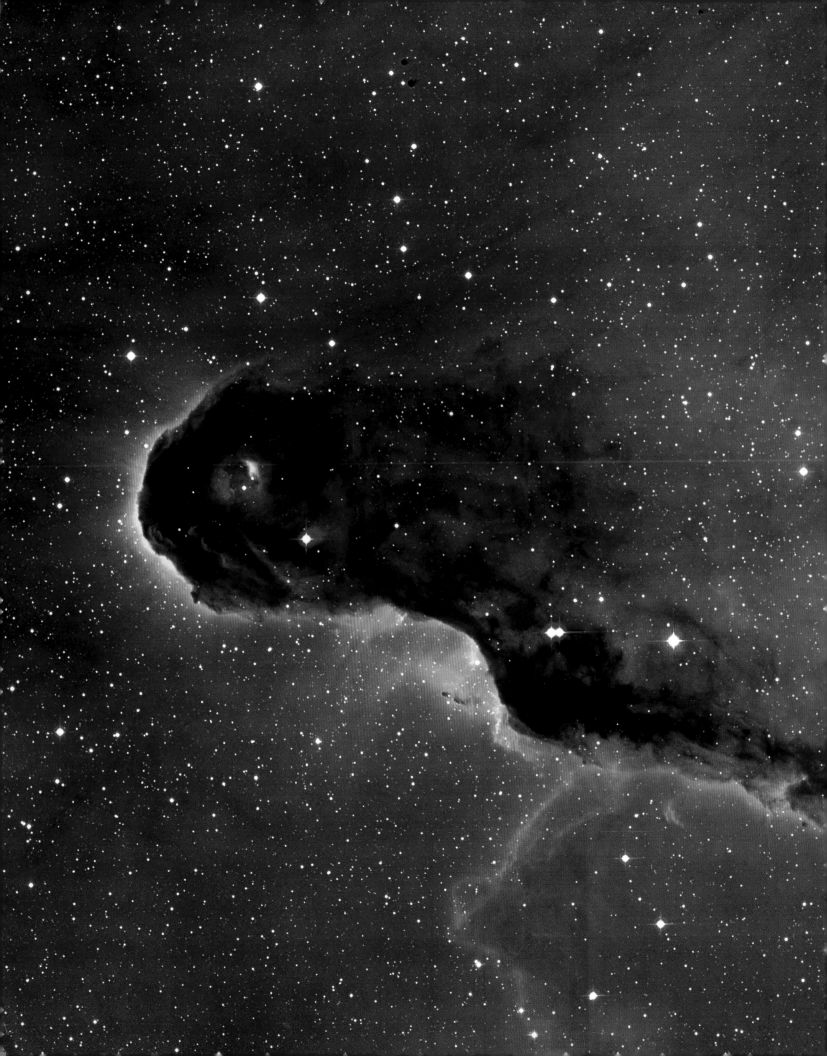

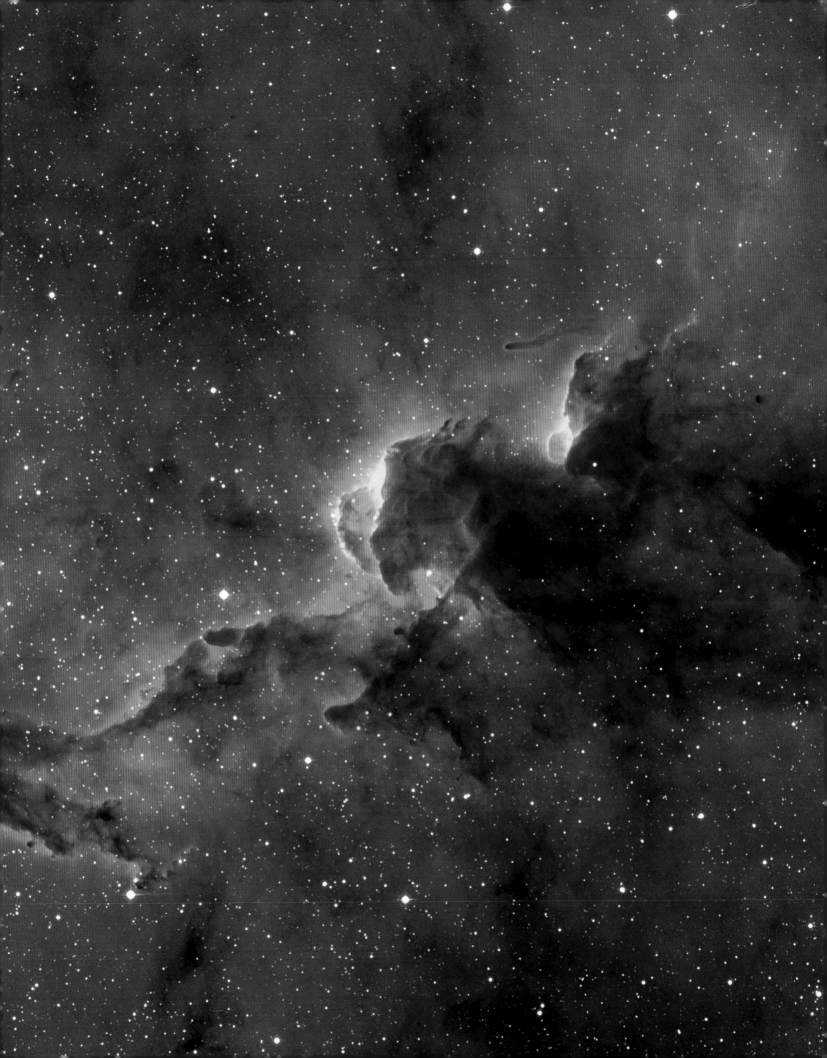

PART 1

EXPLORING THE SOLAR SYSTEM

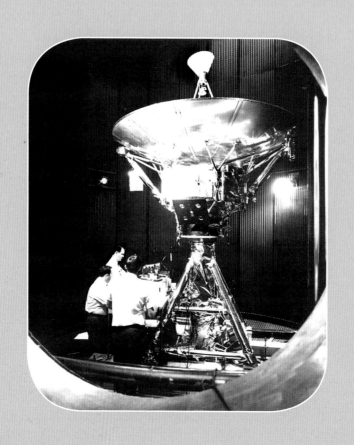

For thousands of years humans have looked at the night-time sky and have tried to make sense of the changes they have seen. The changing shape of the Moon, the changing of the seasons, the different stars visible at different times of the year, the "wandering stars" (planets) which appear to move among the other stars; all these things have puzzled civilisations and have challenged us for an explanation.

The model of our Solar System which has been believed for the longest period of time was one developed by Greco-Egyptian astronomer Ptolemy in the second century AD. It put the Earth at the centre of the Universe, with the Sun, Moon and planets all orbiting us. This model was challenged by Nicolas Copernicus in the mid-sixteenth century when he suggested that the Sun, not the Earth, was at the centre. By the early 1600s Galileo had shown, using a telescope, that Ptolemy was wrong and that the planets did indeed orbit the Sun; the Earth was demoted to being just one of the Sun's family of planets. Galileo found four moons orbiting Jupiter, and in the mid-1600s Christiaan Huygens discovered Saturn's largest moon Titan. With better telescopes Uranus and Neptune were added to the list of previously known planets, and the first asteroids were discovered. The advent of spectroscopy in the mid-1800s allowed us to study the composition of the atmospheres of the planets and photography allowed us to find fainter and fainter objects.

With the birth of the space age in 1957, it was not long before we started sending instruments beyond Earth's orbit to explore our Solar System. In the 1960s and 1970s a suite of space probes visited our nearest neighbours the Moon, Mars and Venus. The Apollo programme sent human beings to the Moon, with twelve walking on the lunar surface between 1969 and 1972. By the late 1970s we had successfully landed probes on the surfaces of Venus and Mars. But, this was just the beginning. In the mid-1970s we started sending space probes to the outer planets. First Pioneer and then Voyager travelled the vast distances to give us our first ever close up images of Jupiter and her moons, Saturn, Uranus and Neptune. These probes continue to travel, leaving our Solar System: the most distant objects ever sent from Earth.

The exploration of our Solar System has continued. The Galileo spacecraft was sent to study Jupiter and her moons in more detail, the Cassini space probe continues to orbit and study Saturn, her ring system and her moons. We have sent several rovers to the Martian surface, these rovers are helping us to understand the geology of Mars and whether it has ever had the conditions necessary for life. In 2015 the Rosetta mission caught up with a comet and dropped the Philae lander to the comet's surface, and the New Horizons spacecraft flew past Pluto showing details of its surface for the first time. In a little over half a century, we have sent dozens of spacecraft to every planet and many of the major moons in our Solar System.

The next few decades will see us send ever increasingly sophisticated spacecraft on ever more ambitious missions. They should also see us send human beings to another planet. It is almost inconceivable that by the end of this century we will not see human beings living on Mars, just one of the next steps in our continuing exploration of our Solar System.

Opposite: Technicians prepare Pioneer F (which became Pioneer 10) spacecraft for testing in the Space Simulation Chamber at TRW Systems, Redondo Beach, California in 1972. Pioneer 10 was the first mission to take a close-up look at the planet Jupiter, its moons and environment.

ECLIPSED MOON OVER THE VERY LARGE TELESCOPE

Digital Single Lens Reflex Camera
Visible Light

- An eclipsed Moon is often a dramatic sight, particularly when it turns red in Earth's shadow. This happens when light from the Sun reaches the Moon by passing through the Earth's atmosphere; the same physics which causes the Sun to appear redder at sunset can cause the Moon to take on this haunting red glow. It is a dramatic sight to see anywhere, but particularly when it is captured above the European Southern Observatory's flagship facility, the Very Large Telescope (VLT).

  In this image, taken on 21 December 2010, the eclipsed Moon is visible as a reddish disk above one of the VLT's four enclosures, Kueyen (also known as UT2, second from left in this image). Also visible is the Milky Way arching overhead, and the large and small Magellanic Clouds. If it were not for the eclipse these would not

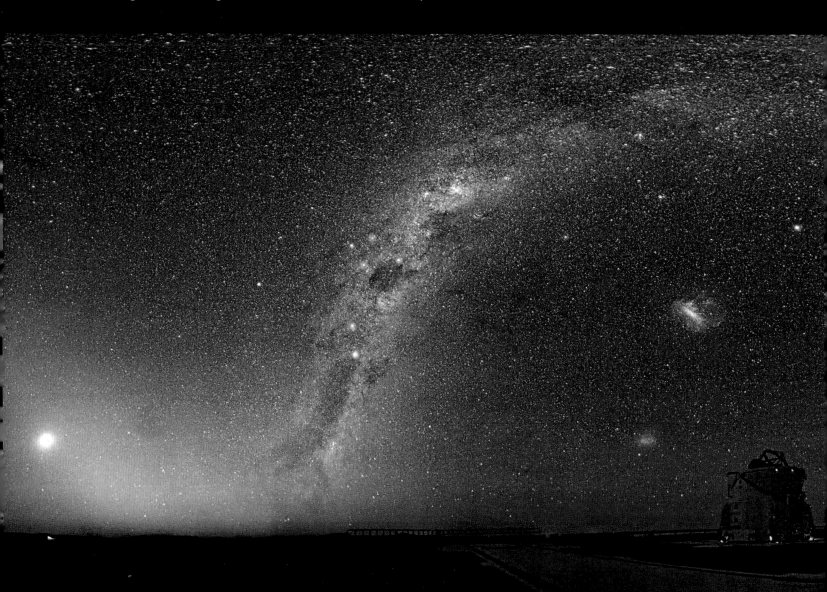

be so easily visible, as the glow of the full Moon would drown out their faint light. Down near the horizon on the left of the image is the bright planet Venus.

The European Southern Observatory (ESO) is a collaboration between 16 European nations and was founded in 1962. The VLT is currently its largest visible light facility, consisting of four 8-metre telescopes which can either operate separately or together to achieve a very high resolution. The VLT is located in Cerro Paranal in the Atacama Desert of northern Chile, at an altitude of 2,635 metres. This location provides more than 340 clear nights a year, one of the best of anywhere in the World.

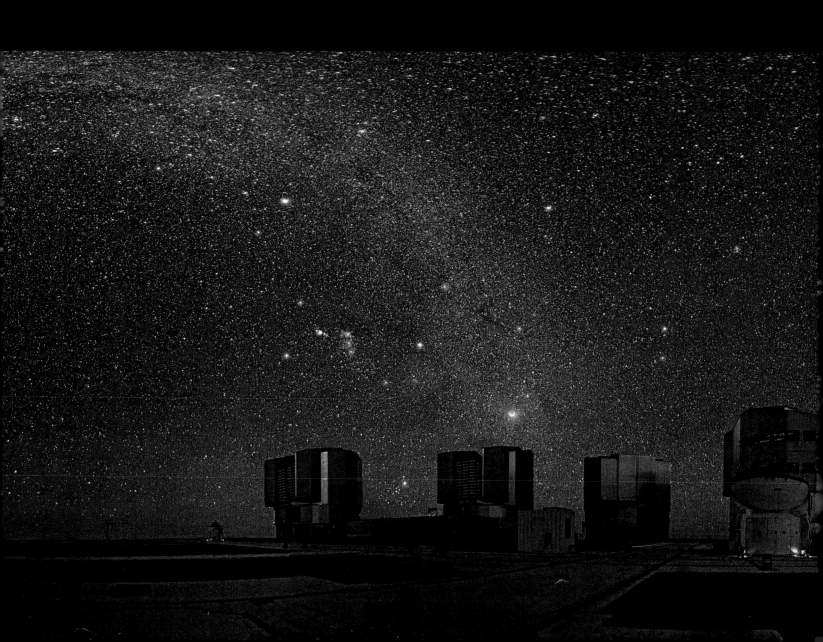

The surface of our Sun can look very calm when viewed in visible light. The only signs of any obvious activity are the well-known sunspots; these dark areas have been studied for centuries and are found to increase and decrease in number on a regular 11-year cycle. In the early part of the twentieth century detailed studies of sunspots showed that they were areas of intense magnetic field on the Sun's surface; their dark appearance is caused by their inhibiting the transport of heat from the interior and thus being cooler than their surroundings. Other evidence that there may be more to the Sun's activity than we normally see is revealed in total solar eclipses. During the brief minutes when the Sun's disk is covered by the Moon a thin atmosphere can be seen extending far out into space; this is the Sun's corona.

The calm appearance of the Sun's surface is shattered when we look at shorter wavelengths. The main picture shows the Sun in soft X-rays, wavelengths which are just a little shorter than ultraviolet light. They were taken by the Yohkoh satellite, a Japanese solar mission which was launched in 1991. In soft X-rays we can see loops of hot gas which are associated with sunspots. These loops (called solar prominences) rise from the solar surface, and in some cases the loops break in what is called a solar flare. The montage was made of images taken between August 1991 and September 2001 showing how the Sun's activity changes over an 11-year cycle.

A particularly strong solar flare is called a coronal mass ejection. These cause charged particles to hurtle out into space; it is these charged particles which cause the aurorae when they enter our atmosphere.

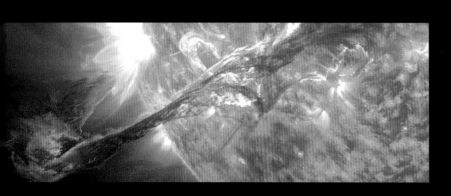

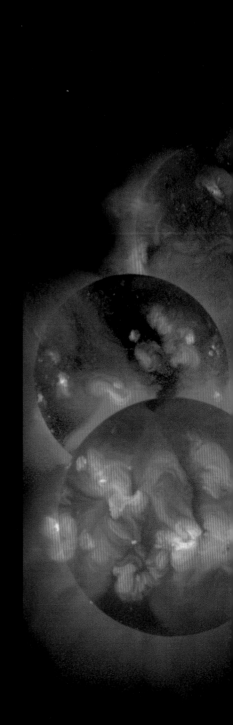

Above: On 31 August 2012 a long filament of solar material erupted, imaged by STEREO and SOHO. The coronal mass ejection, or CME, travelled at more than 1,400 kilometres per second. The CME caused aurorae on the night of 3 September.

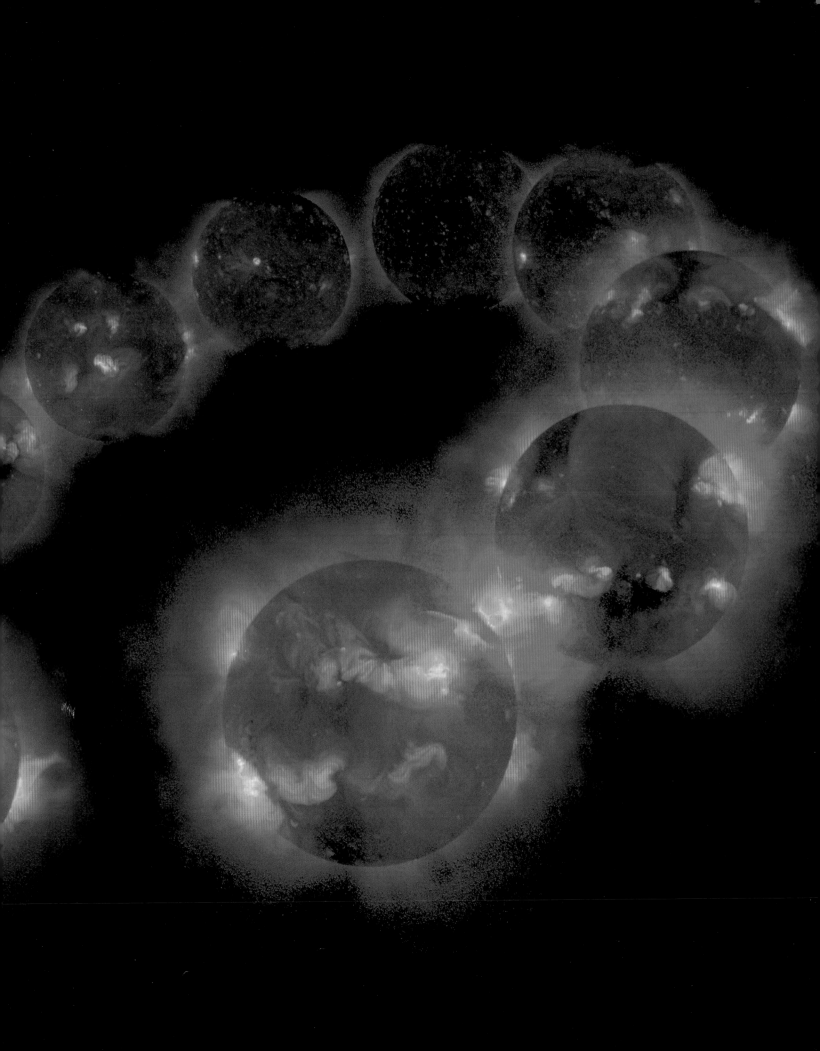

THE SUN'S CORONA

SOlar and Heliospheric Observatory (SOHO)
Ultraviolet Light

- Seeing the Sun's corona during an eclipse is one of the most breathtaking sights we can see during these rare celestial events. Over the centuries, astronomers have travelled to the other side of the Earth to be in the path of totality of a solar eclipse, enabling them to catch a glimpse of the Sun's outer atmosphere which normally remains hidden from view. From the ground however, even if we block out the light of the Sun by placing a disk over it (called a coronagraph), the scattering of light in the Earth's atmosphere means that the faint corona still remains invisible to us. The same is not true in space as, in the absence of air, no such scattering of light occurs. Thus, satellites in space can place a coronograph over the Sun's disk which allows us to study the corona all the time, even when there is no eclipse.

  The image of the Sun's corona shown here was taken by the SOlar and Heliospheric Observatory (SOHO), a joint NASA and ESA satellite which began observing the Sun in 1996 and continues to operate. SOHO lies in an orbit around the Sun at a special point between the Earth and the Sun known as the Lagrangian-1 point (L1). Normally, objects closer to the Sun than the Earth will orbit it more quickly, Venus and Mercury are good examples. However, L1 is a point in space where the gravitational pull of the Sun and that of the Earth are balanced, allowing the satellite to orbit closer to the Sun than the Earth's orbit but taking the same 1-year period to complete its orbit.

  Taken in May 2000, this SOHO image clearly shows the Sun's outer corona stretching out into space. The blue disk at the centre is the coronagraph used to block out the Sun's direct light, and the white circle shows the extent of the Sun's disk. The image also shows Mercury, Venus, Jupiter and Saturn, which would normally be hidden from view, lost in the glow of the Sun. The star cluster Pleiades (see page 85) is also visible in the upper-left of the image. Satellites like SOHO have greatly enhanced our understanding of the Sun's corona and how it varies on a day by day basis.

Opposite above: In this SOHO image referred to in the text, Mercury is the bright planet at left, beneath the Pleiades cluster. Venus is at far right, with Jupiter just below and to the left of it, with Saturn further left again, below the centre line of the picture.

Opposite below: The total eclipse of March 2006, as seen from the International Space Station.

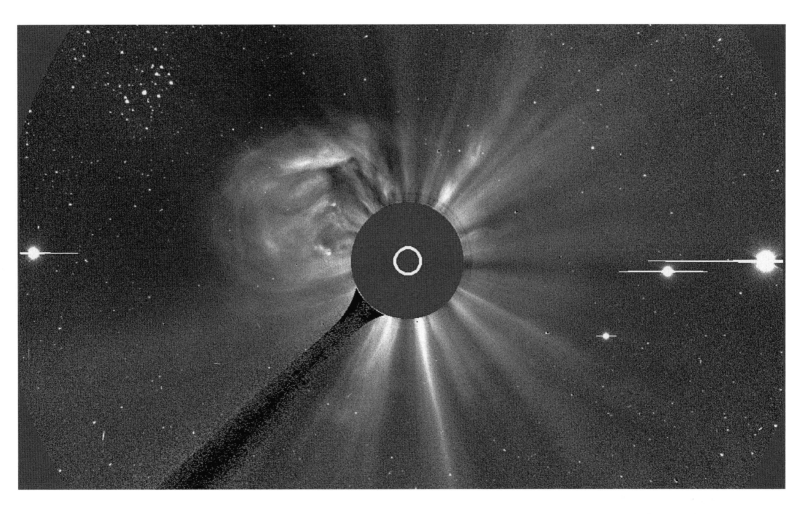

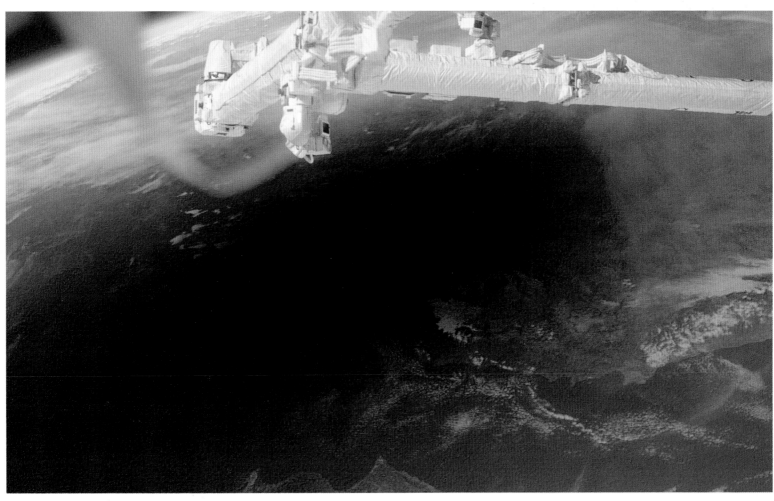

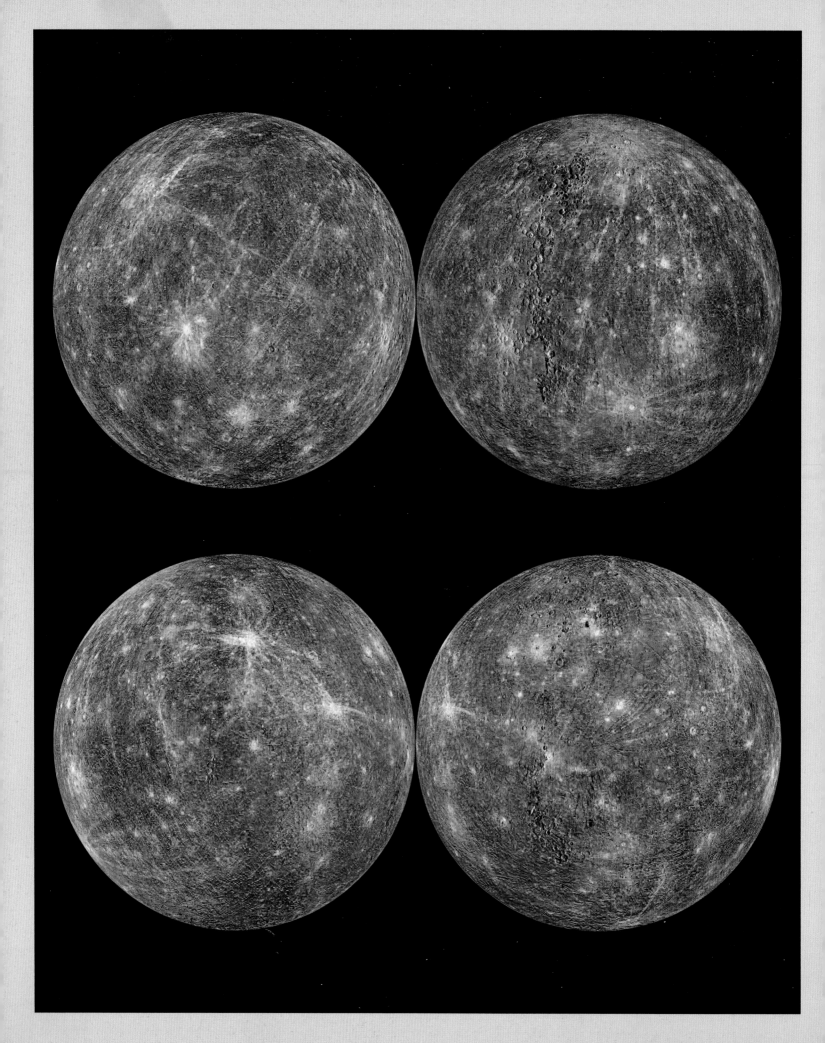

MERCURY BY MESSENGER

MESSENGER Space Probe
Visible Light

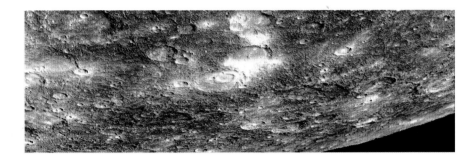

Mercury, the closest planet to the Sun, is the least explored of the five naked-eye planets. To date, we have sent only two space probes to the planet; Mariner 10 in the mid-1970s and more recently MESSENGER, which first arrived at Mercury in January 2008. Although it comes much closer to Earth than any planet other than Venus and Mars, sending a space probe to Mercury poses a significant challenge, due to the gravity of the Sun. As a spacecraft heads towards Mercury it speeds up due to the Sun's gravitational pull, just as if we were rolling a ball down a hill. Thus, the challenge is to slow the spacecraft down enough so that it does more than rapidly whizz past Mercury. As Mercury is also a small planet, it does not have a strong enough gravitational pull of its own to slow down a space probe, making it difficult to put a probe in orbit about it.

Mariner 10 made three close approaches to Mercury, passing within 330 kilometres of its surface, but it did not orbit it. MESSENGER also made three fly-bys of Mercury, in January 2008, October 2008 and September 2009. It was then put into an elliptical orbit around the planet in March 2011. It mapped the surface of Mercury for three years, but by April 2015 its fuel was expended and it crashed into the surface on April 30, 2015.

The images shown here are enhanced spectral scans of the two hemispheres of Mercury as imaged by MESSENGER's Mercury Atmospheric and Surface Composition Spectrometer (MASCS). MESSENGER discovered organic compounds in the surface of Mercury, and water ice which resides inside craters near the north pole which never receive any sunlight. It also found evidence for a liquid iron core, past volcanic activity on the planet's surface, and of large amounts of water in Mercury's very thin atmosphere. In January 2017, ESA and the Japanese space agency will launch BepiColombo, two joint satellites which are due to begin orbiting Mercury in January 2024 and spend at least one year studying details of the planet's surface, thin atmosphere and interior composition.

Above: The yellow regions in this MESSENGER image of the southern hemisphere of Mercury are pyroclastic vents stretching from roughly -60 degrees latitude, in the crater Hesiod, to about -51 degrees latitude. These vents are thought to stem from explosive eruptions powered by volcanic gases.

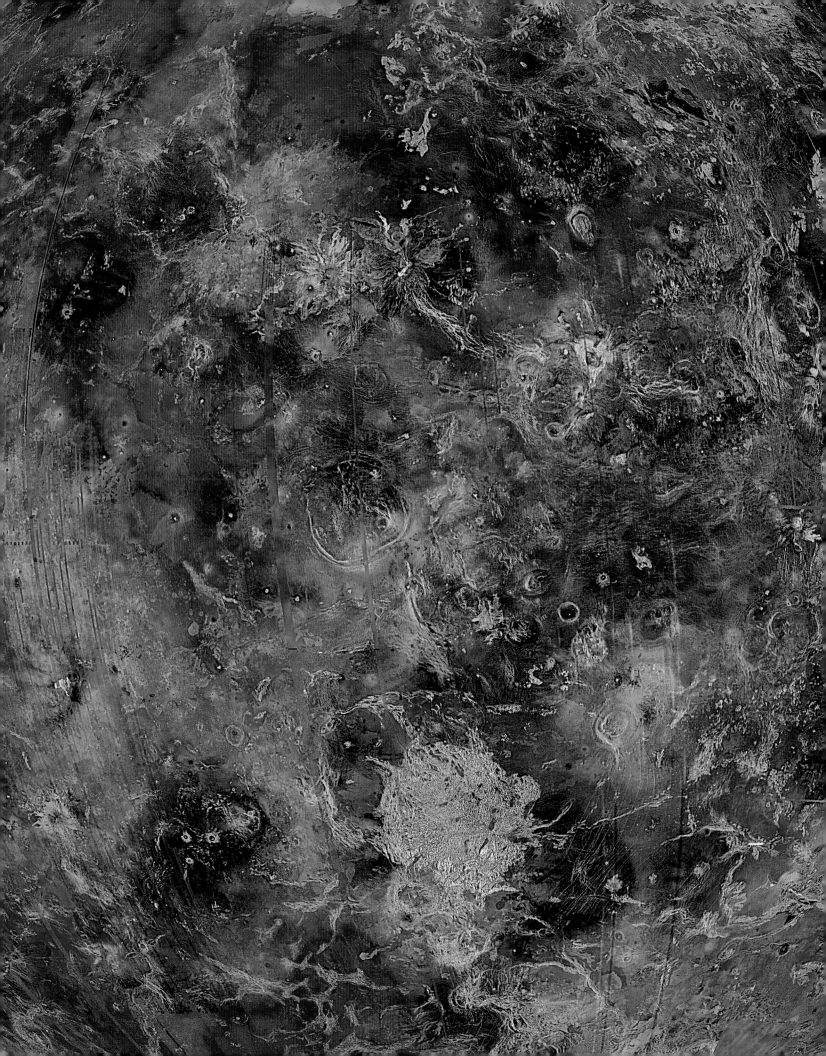

VENUS

Magellan Spacecraft and Arecibo Radio Telescope
Radio Waves

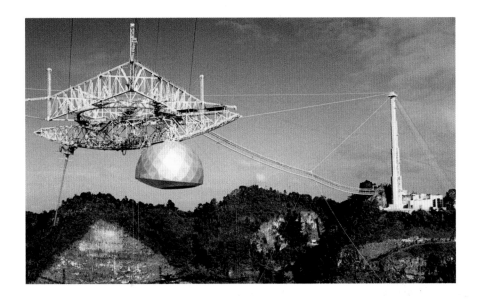

From Earth the surface of Venus, long thought of as our twin, is hidden from view by a thick atmosphere. For centuries astronomers had strained to see the surface, imagining a lush paradise similar to the Earth's tropics. The Soviet Union's Venera 1 was sent to Venus in February 1961, but failed to reach its destination. The USA's Mariner 2 successfully flew past Venus the following year, and measured a temperature in excess of 450 Celsius.

A succession of Soviet and American probes continued to visit Venus throughout the 1960s. In 1967 Venera 4 became the first probe to enter the atmosphere of another planet; it measured the atmosphere to be 95 percent carbon dioxide and the atmospheric pressure to be about one hundred times greater than Earth's. Venera 7 landed on the surface in December 1970 and measured the surface temperature to be between 455 and 475 Celsius. Later probes showed sulphuric acid droplets in the upper atmosphere.

With such a thick atmosphere, seeing the surface of Venus in visible light is not possible. But, in May 1989 NASA launched the Magellan spacecraft, which would use radio waves to peer through the impenetrable clouds and "see" the surface of our closest planet for the first time. Over a four-year mission, Magellan mapped 98 percent of the surface. It found no evidence for plate tectonics, and the lack of impact craters suggests that the surface is relatively young. Magellan also saw long lava channels, some thousands of kilometres in length. The radar map here is a mosaic of observations by the Magellan spacecraft, with gaps in the mapping filled by images from the Arecibo Telescope on Earth.

Above: The Arecibo Radio Telescope in Puerto Rico fills a natural crater in the rainforest. The concave radio dish is 300 metres in diameter and is made of thousands of aluminium panels to focus incoming radio waves onto a 9-tonne platform which is suspended above it.

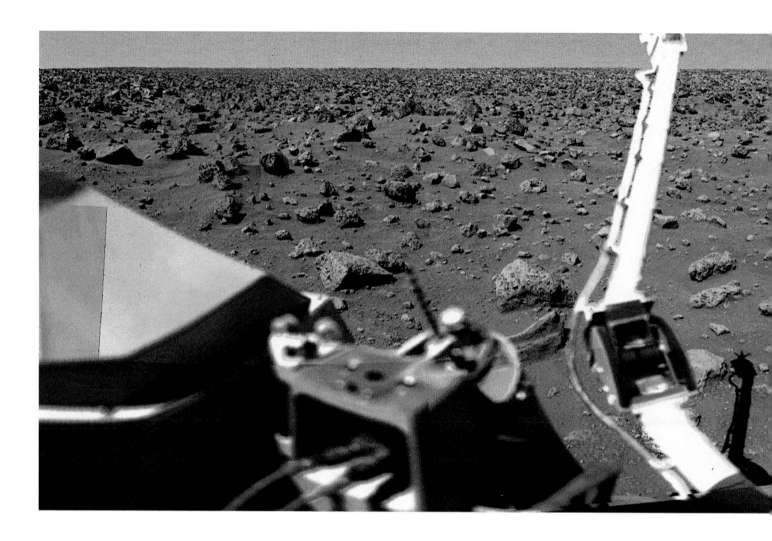

VIKING 1 IMAGES OF THE MARTIAN SURFACE

Viking 1
Visible Light

- In August and September 1975 NASA sent two probes to land on the surface of Mars. Viking 1 became the first probe to successfully touch down on the Martian surface, landing on 20 July 1976. It was followed a few months later when Viking 2 successfully landed on 3 September. Viking 1 touched down in the western Chryse Planitia ("Golden Plain"), a little less than 23 degrees north of the Martian equator. Viking 2 landed on the other side of Mars at Utopia Planitia, about 200 kilometres to the west of the crater Mie at a latitude of just over 48 degrees north.

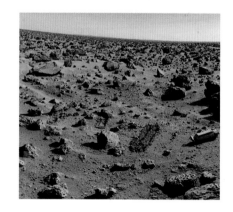

Each probe was transported to Mars by its own orbiter, but the orbiters were not merely transport vehicles. Each orbiter also contained scientific instruments to measure the structure and composition of the Martian atmosphere, and imaging cameras allowing them to obtain high resolution images of possible landing sites

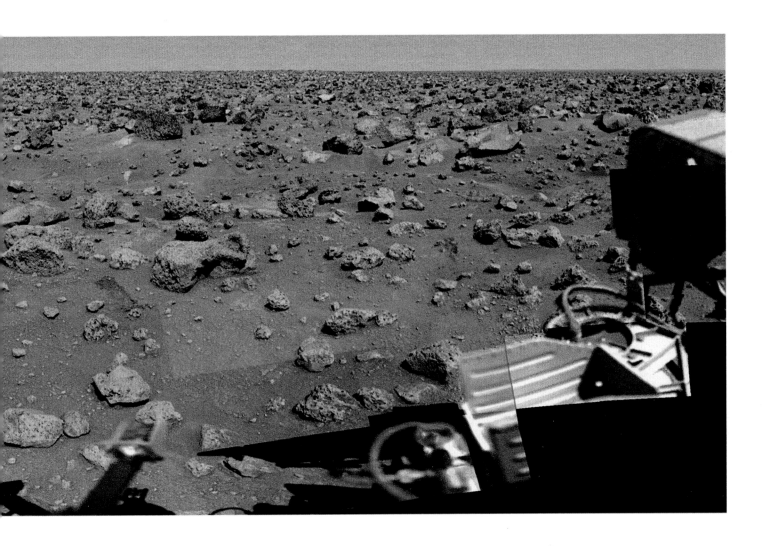

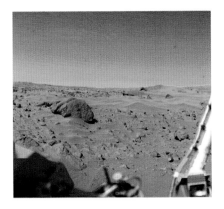

before the landers separated. Once the landers were on the surface the orbiters continued their own scientific programme, and also relayed the data being taken by the landers back to Earth. The images shown here are the first clear images ever transmitted from the surface of Mars.

In addition to the many rocks and boulders visible in these Viking 1 images, we can also see the trenches dug by the soil sampling device. The soil was analyzed for signs of life; the results were inconclusive and past or current life on Mars still remains an open question.

THE SURFACE OF MARS

Mars Curiosity Rover
Visible Light

- Since it landed on the surface of Mars on 6 August 2012, Mars Curiosity Rover, part of NASA's Mars Science Laboratory mission, has travelled more than 11 linear kilometres across the surface of Mars taking samples to investigate, among other goals, whether Mars has ever provided conditions capable of sustaining microbial life.

    Curiosity began its journey in the Gale crater, and is now in the foothills of Mount Sharp. It is the largest and most sophisticated rover that we have ever put on the Martian surface. It is armed with a suite of instruments which enable it to study the geology of Mars, including stereoscopic cameras, spectrographs, and even drills and an ability to take samples of soil and rock and bake them in an oven and study the gases released.

    Curiosity has 17 cameras on it, the most of any NASA planetary mission ever. The Mars Descent Imager took pictures as the rover was landing on Mars. Then there is a camera mounted on the end of the arm which takes close-up, high-resolution colour photos. This was used to take the "selfie" shown on the following pages.

    At ground level we have the hazard avoidance cameras, or the HazCams. There are four of these up front, and four at the back, and these are used to take pictures in 3D of the terrain near the wheels. On the mast there are the cameras that take most of the pictures for the mission. There are also navigation cameras, which take pictures that are used to drive the rover, and other mast cameras are colour imagers, which are used to do geology investigations.

    Finally, there is the remote microscopic imager, part of the ChemCam laser instrument, and that is used to document the laser spots that the rover makes on the surface.

Right: Victoria Crater, in which the Mars Opportunity Rover spent some time, imaged by NASA's Mars Reconnaisance Orbiter.

Below: The "Kimberley" formation near the base of Mount Sharp. The colours are adjusted so that rocks look approximately as they would if they were on Earth, to help geologists interpret them. This "white balancing" to adjust for the lighting on Mars overly compensates for the absence of blue on Mars, making the sky appear light blue and sometimes giving dark, black rocks a blue cast.

Following pages: A Martian selfie taken from the end of Mars Curiosity Rover's robotic arm. Seen here at the "Mojave" site, where its drill collected the second sample from Mount Sharp. This scene combines dozens of images it took during January 2015.

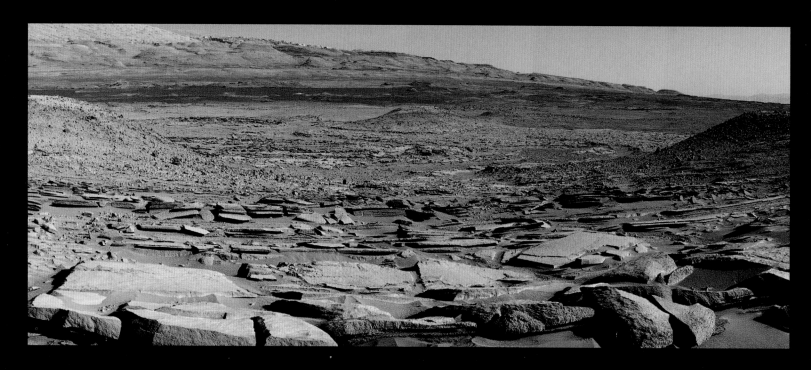

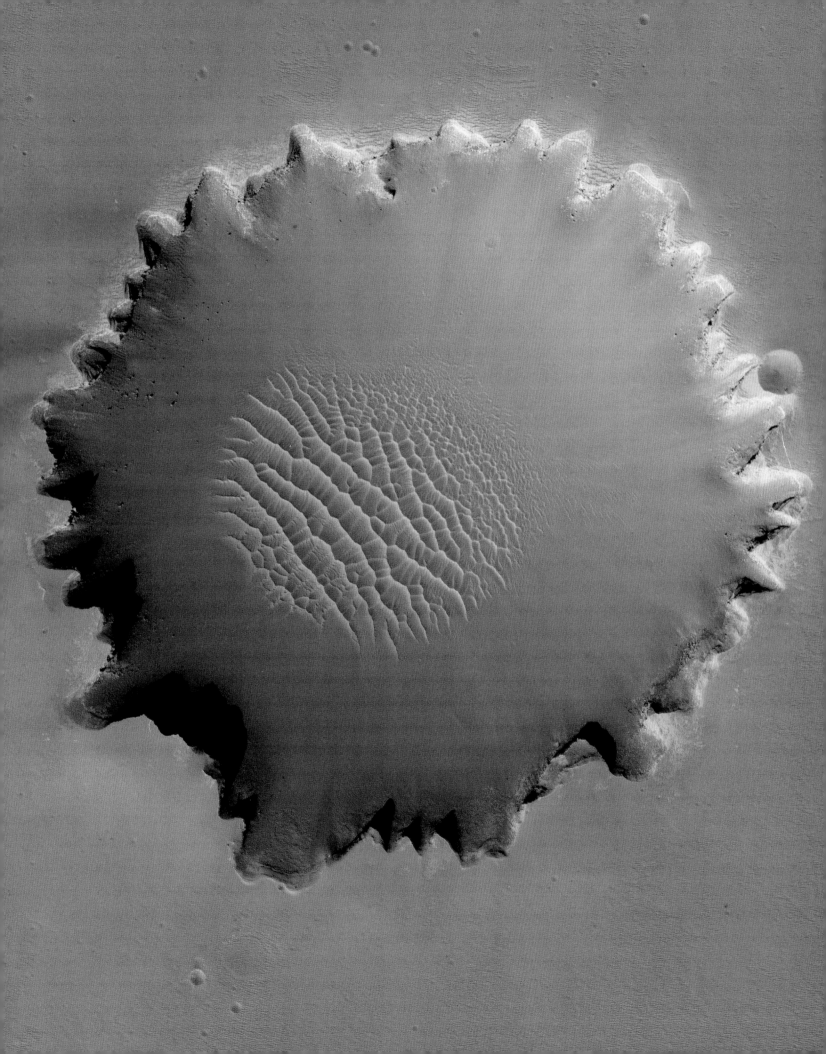

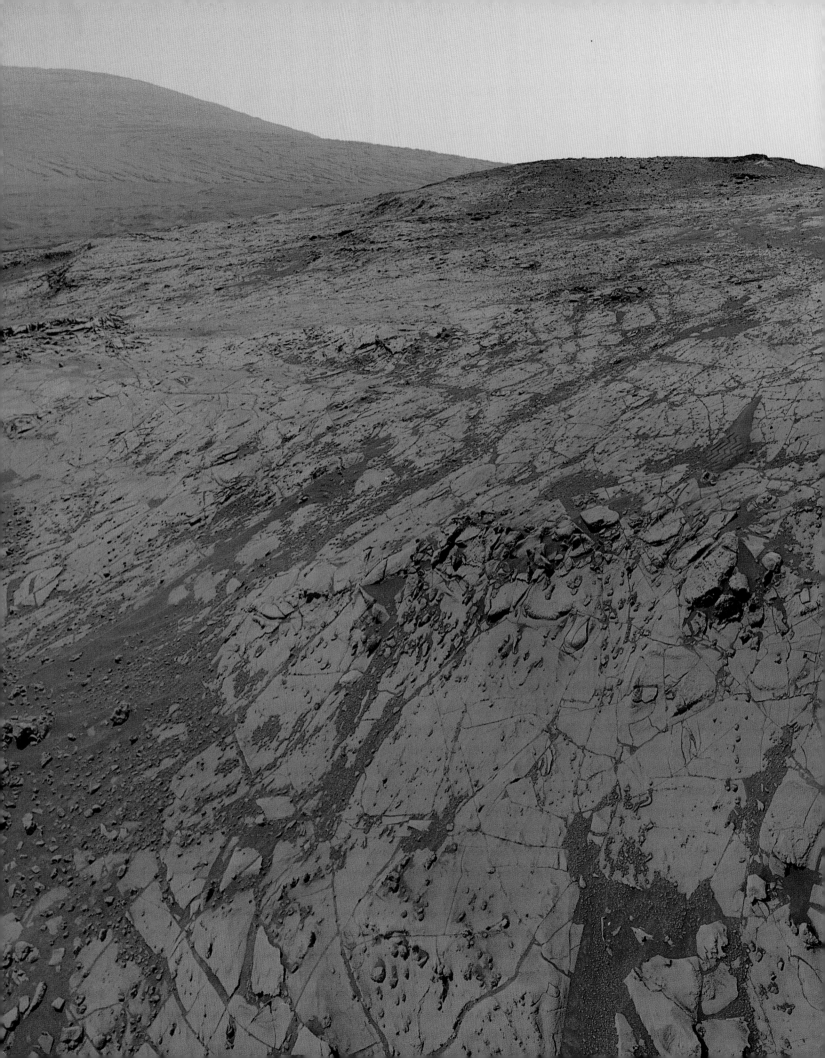

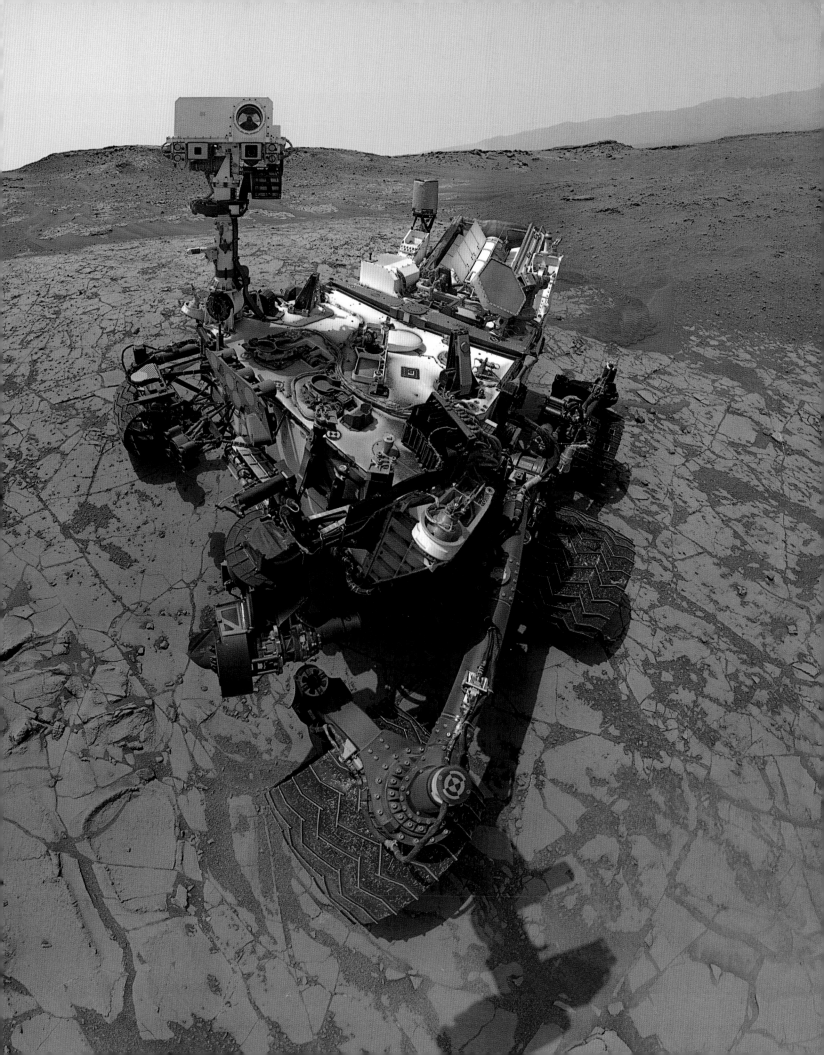

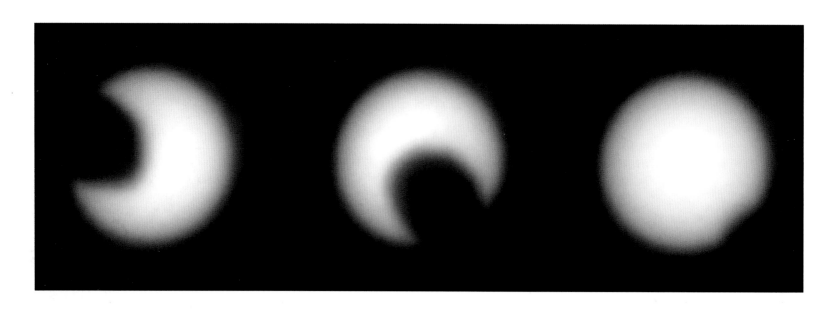

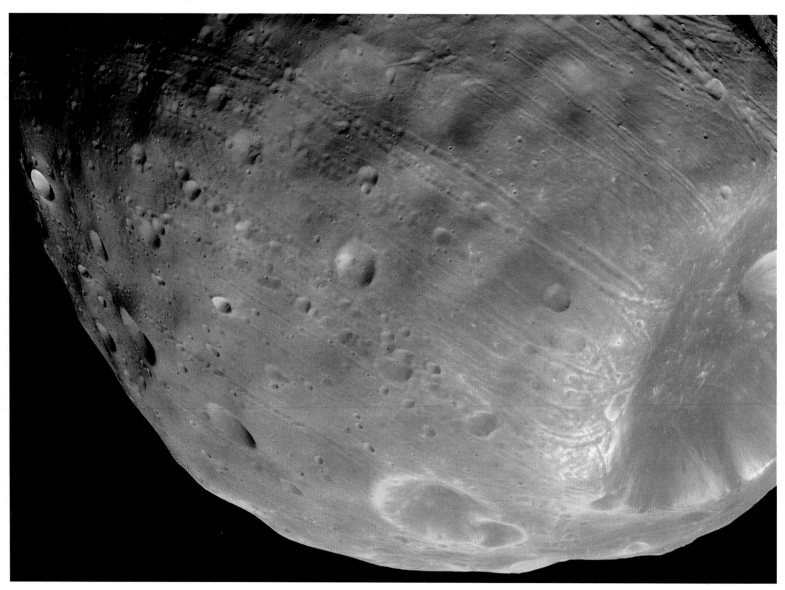

TRANSIT OF PHOBOS

Mars Opportunity Rover
Visible Light

• Mars Opportunity Rover took this dramatic image sequence of Phobos (top left), one of the two moons of Mars, passing in front of the Sun. It was taken in March 2004, early in Opportunity's mission, and shows that any cameras which we have placed on the surface of Mars can do more than just study the Martian surface. Unlike with our Moon, Phobos is not able to hide the whole disk of the Sun and cause a total eclipse. Instead it transits the Sun in much the same way as Venus or Mercury transit the Sun as seen from Earth.

Whereas transits of Venus or Mercury from Earth are quite rare (a few each century), and solar eclipses due to our Moon happen roughly twice a year, the geometry of Phobos's orbit make its transits across the Sun much more common. Mars rotates once on its axis every 24 hours and 37 minutes, but Phobos' orbit is so close to the Martian surface that the moon takes only 7.6 hours to complete one orbit. In addition, it orbits close to the equator of Mars, and this means that transits occur somewhere on Mars most days. Several of these have been captured by various rovers, but the one shown here was the first to be seen.

Phobos is the closer of the two moons. Deimos, the other moon, is roughly three times further away and takes some 30 hours to orbit Mars. Both were discovered in 1877 by Asaph Hall, an astronomer at the US Naval Observatory. Given their appearance, size and composition they were probably captured from the nearby asteroid belt.

Opposite below: The High Resolution Imaging Science Experiment (HiRISE) camera on NASA's Mars Reconnaissance Orbiter took this image of Phobos from a distance of about 6,800 kilometres. Stickney crater is at lower right.

Below: Deimos imaged by the HiRISE camera on NASA's Mars Reconnaissance Orbiter.

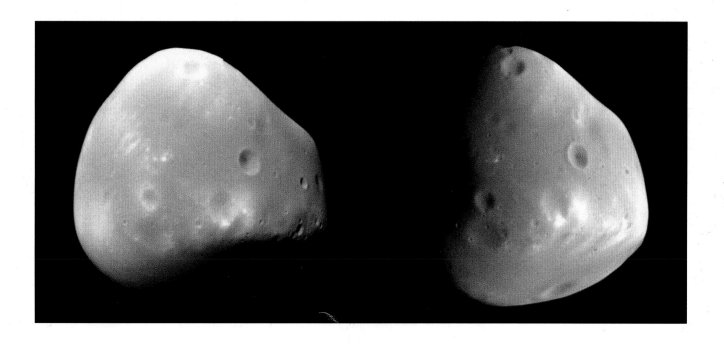

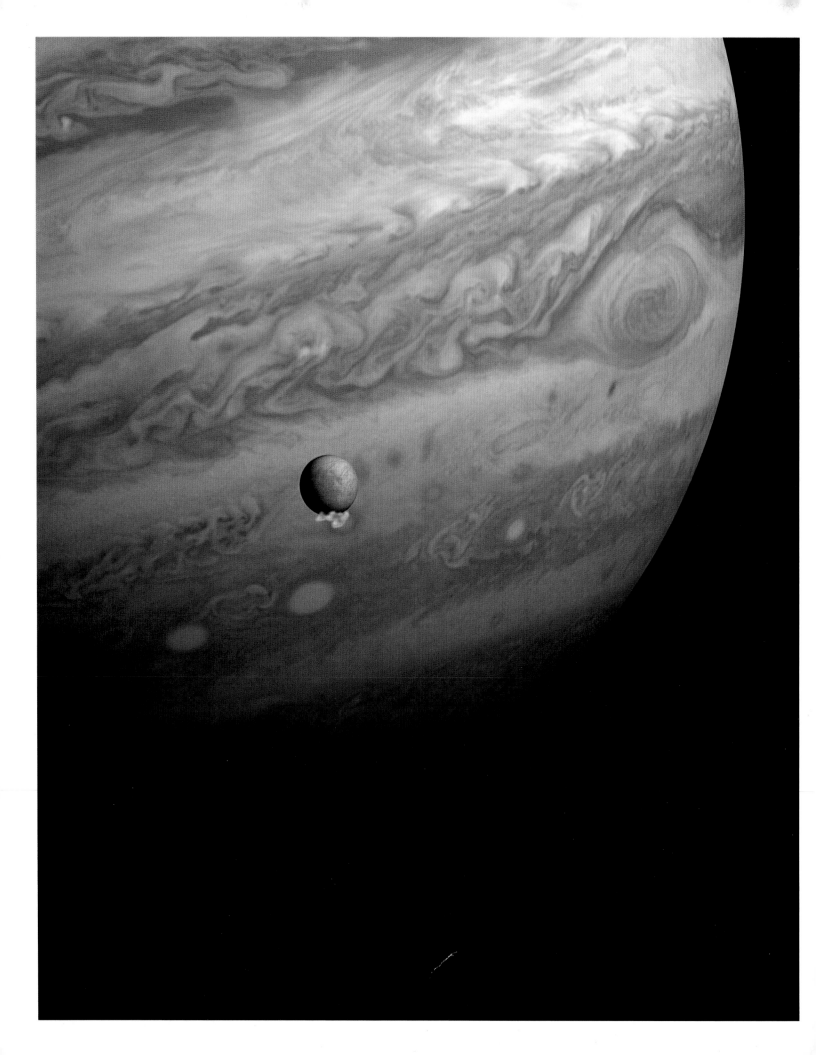

PIONEER 10 AND GALILEO AT JUPITER

Pioneer 10 and Galileo Space Probes and Hubble Space Telescope
Visible Light

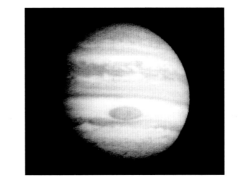

- In December 1973 the space probe Pioneer 10 took the first ever close-up images of the planet Jupiter (right). Never before had such detailed pictures of the Solar System's largest planet been seen. The images taken by Pioneer 10 became iconic; they represented our first foray into the outer Solar System. The great red spot, the giant storm system first seen by telescopes in the mid-1600s, is clearly visible in this image taken from less than three million kilometres away.

  Pioneer 10 was launched in March 1972 and became the first space probe to traverse the asteroid belt and travel on to Jupiter. It eventually went on to pass the orbits of Saturn in 1976, Uranus in 1979 and Neptune in 1983, although it did not visit any of these planets.

  Pioneer 10 was followed in April 1973 by Pioneer 11 which also visited Jupiter, in late 1974, but then went on to visit Saturn in September 1979. The two probes taught us a great deal about the environments of these two giant planets, and also served as a precursor to the much more ambitious Voyager 1 and 2 probes which were launched in 1977. Today, Pioneer and Voyager are hurtling out of the Solar System, passing the point where the influence of the Sun's radiation is less than the radiation from other, nearby stars.

  The images below were taken by the Galileo space probe, which arrived at Jupiter in 1995 to orbit the planet and study its large moons.

Opposite: Images of Jupiter and Europa from the Hubble Space Telescope, with an artist's impression of water vapour plumes emitted from Jupiter's moon Europa.

Below: Jupiter's moon Ganymede seen by Galileo's Near-Infrared Mapping System showing water ice in green (centre) and (right) the locations of minerals in red and the size of ice grains in blue. The left image is from Voyager.

Above: This view of Jupiter shows the giant planet's cloud tops taken by the Pioneer 10 spacecraft as it flew past Jupiter some 2,695,000 kilometres away. It shows the 40,000-km long Great Red Spot.

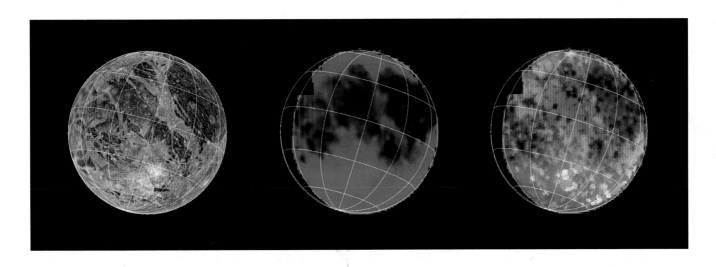

JUPITER'S AURORAE

Hubble Space Telescope
Ultraviolet Light

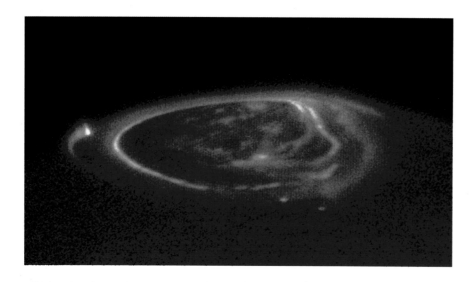

• Jupiter's magnetic field is 14 times stronger than the Earth's, which makes it the most powerful magnetic field anywhere in the Solar System except for sunspots. This strong magnetic field has a number of effects on Jupiter and its closest moon Io. The magnetic field itself is believed to be generated in a similar way to the Earth's, the circulation of an electrically conducting material in a liquid outer core. In the case of the Earth, this molten outer core is iron and nickel, but in Jupiter it is liquid metallic hydrogen (a type of hydrogen which can form under intense pressure). Due to Jupiter's large size and its fast rotation of only 10 hours, the fastest of any planet in the Solar System, this generates the strong magnetic field observed.

The influence of Jupiter's magnetic field stretches a long way out into space. All of Jupiter's large moons (the Galilean moons) orbit within its influence. In the direction away from the Sun it stretches out as far as Saturn's orbit. Volcanic eruptions on Io (see page 36) eject large amounts of sulphur dioxide gas into space, and this gas gets ionized by Jupiter's magnetic field, loading the magnetic field with plasma (ionized gas). The plasma rains down on Jupiter's magnetic poles, and as the charged gas interacts with Jupiter's upper atmosphere it creates permanent aurorae at the poles.

The spectacular image shown here was taken by the Hubble Space Telescope. It shows Jupiter's aurorae at both poles, radiating in ultraviolet light.

Above: A close up of the north polar aurora seen by Hubble.

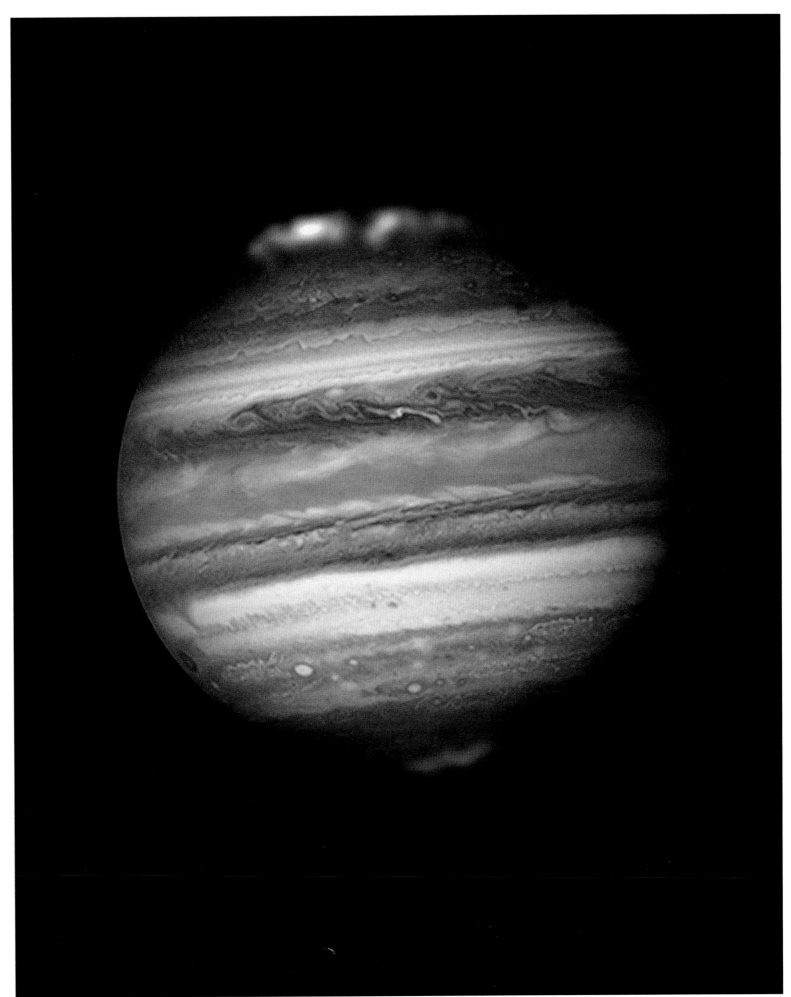

VOLCANOES ON IO

Galileo and New Horizons Space Probes
Visible Light

- Io, the closest moon to Jupiter, is the most volcanically active body in the Solar System. When Voyager 1 flew past in March 1979 it revealed a multi-coloured surface devoid of impact craters. Its surface looked more like a pizza than the surface of a moon. Shortly afterwards Linda A. Morabito, a Voyager navigation engineer, noticed a plume coming from the surface in one of the images. Further analysis of other Voyager 1 images showed nine such plumes, proving that Io was volcanically active.

  The image shown here was taken by the Galileo space probe in June 1997, and the plume from an active volcano can clearly be seen on the limb of Io. Given Io's size, any heat left over from its formation should have long since dissipated. One would also not expect the heat from the decay of radioactive elements in the core of the moon to cause such a level of volcanic activity, so what is the source of its

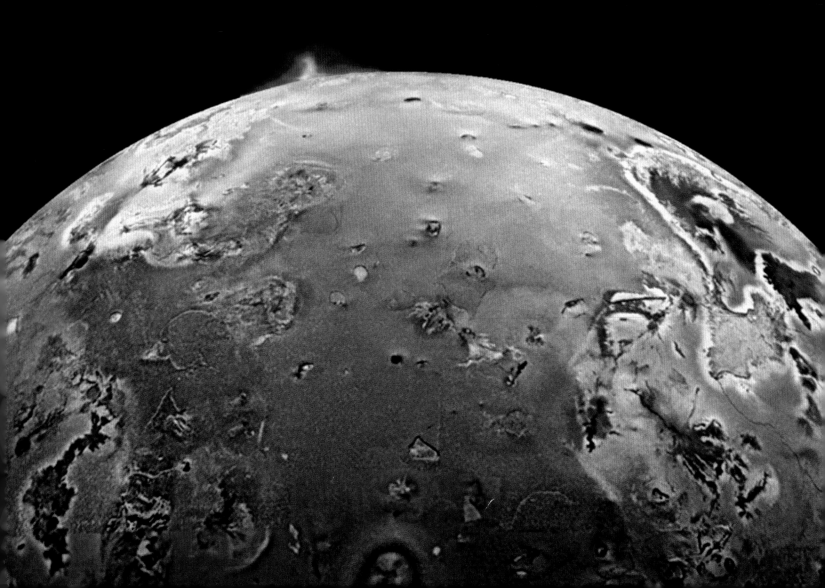

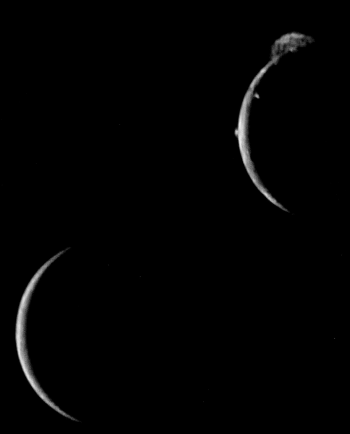

Above: The New Horizons spacecraft imaged Io and Europa, another Jovian moon, in February 2007. A plume from a huge volcanic eruption can be seen at Io's north pole.

interior heat? The answer is to do with tides, the same phenomenon that causes the twice-daily rise and fall of the oceans on the Earth's surface.

Technically, the term tides refers to what happens when the gravitational pull on an extended body differs in different parts of it. Io orbits Jupiter in an ellipse, and so its distance from its parent planet varies. This causes Io to be squeezed and stretched by tidal forces as it rotates. Just like kneading a tennis ball, the constant kneading of Io causes its interior to get so hot that it is molten, and this is the source of its extreme volcanic activity.

EUROPA - HOME TO LIFE?

Magellan Spacecraft and Hubble Space Telescope
Visible Light

- Of all the objects in the Solar System, Jupiter's moon Europa is top of most lists when it comes to places to look for life beyond our Earth. This is because there is strong evidence that it harbours an ocean of liquid water below its icy surface. Europa is one of the four Galilean moons, discovered by Galileo in January 1610. It is the second closest of the four to Jupiter, and takes a little over 3.5 days to orbit its parent planet. When Voyager 1 and 2 flew past the Jovian system in 1979, they provided detailed images of Europa's surface.

  Interpretations of the images, including how the surface changes appearance, provided hints that Europa may have a liquid ocean below its surface. NASA decided to develop a space probe to go into orbit about Jupiter and study it and its moons in far more detail. The Galileo space probe was launched in October 1989 and arrived

Opposite: Three of Jupiter's moons cross the face of the planet: Europa at lower left, Callisto above and to the right of it. Io orbits closer to Jupiter and is approaching Jupiter's eastern limb.

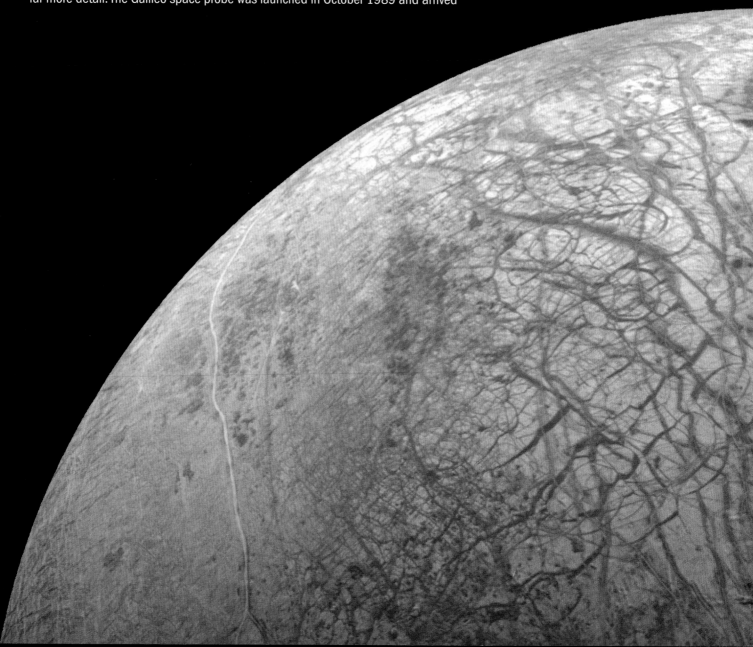

at Jupiter in December 1995. It spent the following eight years examining the Jovian system in unprecedented detail, including numerous close fly-bys of Europa. Its cameras took the images of Europa shown here; the surface is criss-crossed with features called lineae which can change over a matter of weeks.

It is argued that these lineae have been produced by a series of eruptions of warm ice from warmer water beneath. Passing radar signals from the space probe back to Earth which grazed the limb of the moon have also shown evidence for a layer of liquid water. In the future we hope to send a space probe to drill down through the ice and see whether there are any signs of life in this ocean.

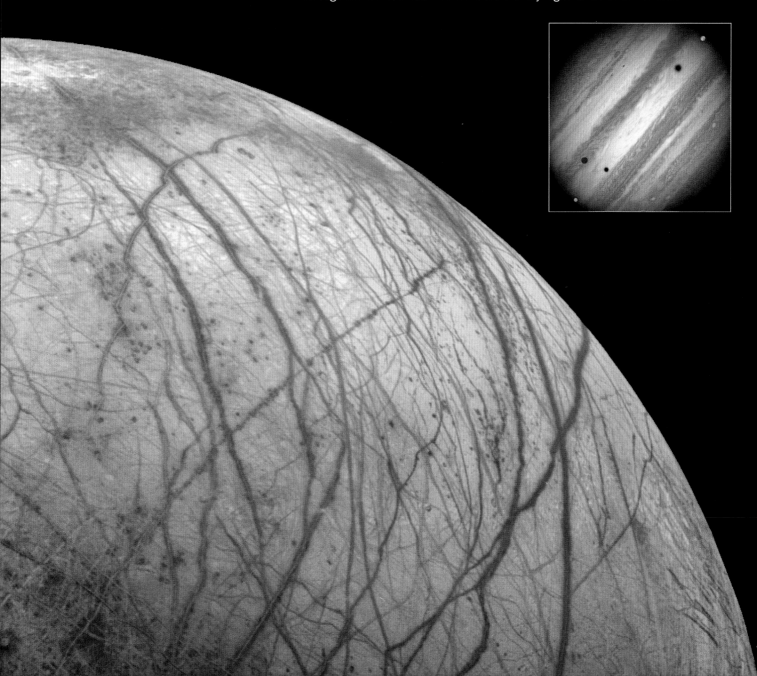

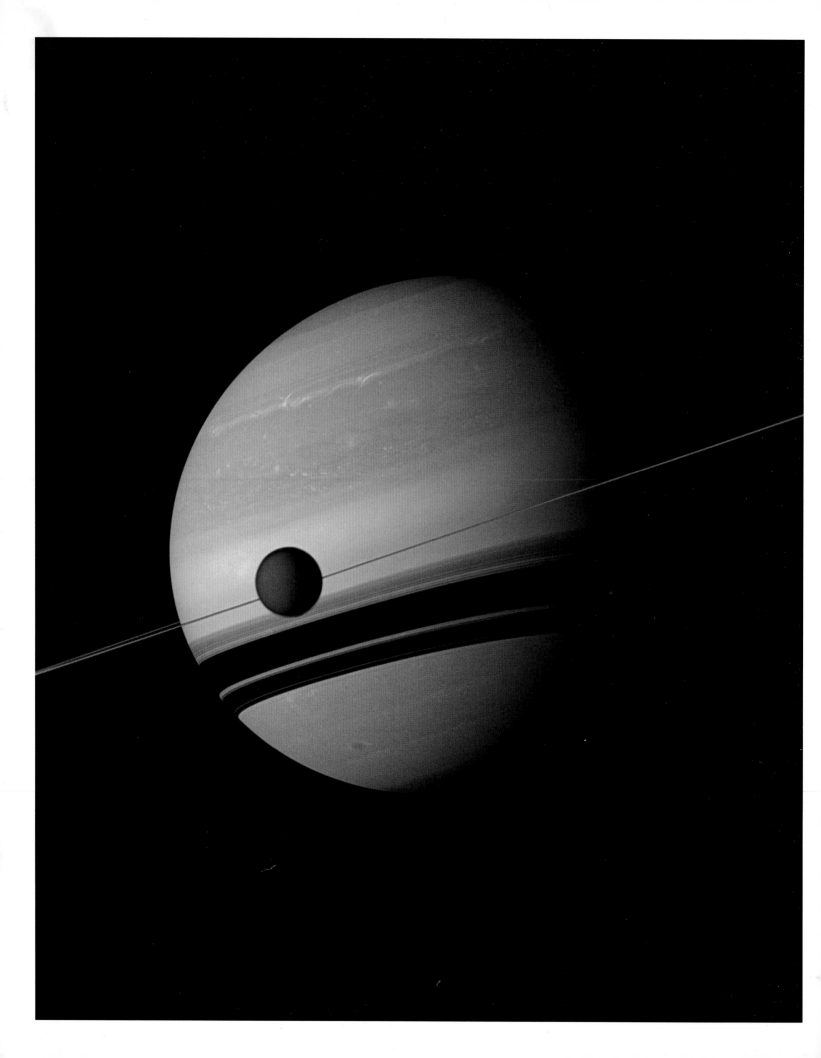

TITAN'S SHORELINE

Cassini and Huygens Probes
Visible Light

- When the Cassini space probe was sent to study the Saturnian system, the decision was made to piggyback a smaller probe, named Huygens, which would descend through Titan's atmosphere and land on its surface. During its descent it took pictures and measured the composition, temperature and pressure of the atmosphere as a function of altitude down to the surface.

As the Huygens probe descended towards Titan's surface on 14 January 2005, it captured the stunning image (below) of features on Saturn's largest moon. Titan was discovered in 1655 by Dutch astronomer Christiaan Huygens, after whom the probe was named. Titan is easily visible through a small telescope; it takes a little under 16 days to orbit Saturn and lies about 10 times further from the planet than the outer edge of the rings. Titan is the only moon in the Solar System which has a significant atmosphere; and most intriguingly this atmosphere is known to be composed of gases that include hydrocarbons, particularly methane. It was another Dutch astronomer, Gerard Kuiper, who in the 1940s was the first to confirm Titan's atmosphere; by looking at the spectrum of the light reflected from it he was able to show that it had an atmosphere which contained methane.

The image shown here is of what appears to be hills, a shoreline, rivers and a plain. It is believed that the hills are composed of water ice, while liquid methane flows in the rivers down to the plain. It is thought that methane can exist as a solid, liquid or vapour on Titan in the same way that water does on Earth.

Opposite: Titan transiting the face of the gas giant Saturn, its parent planet, imaged by the Cassini spacecraft. The rings are aligned in a thin plane behind Titan, and their shadows can be seen on the surface of the Southern Hemisphere of Saturn.

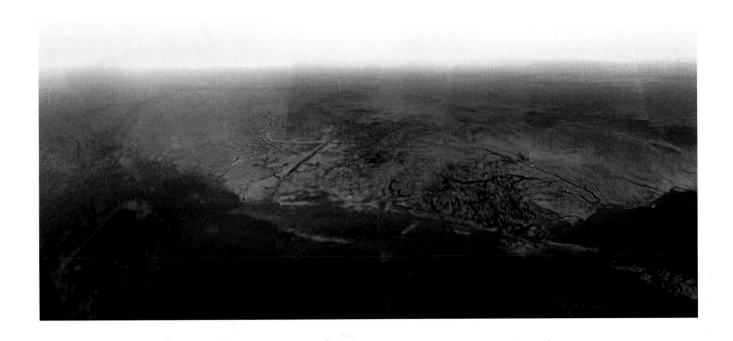

EARTH THROUGH SATURN'S RINGS

Cassini Space Probe
Visible Light

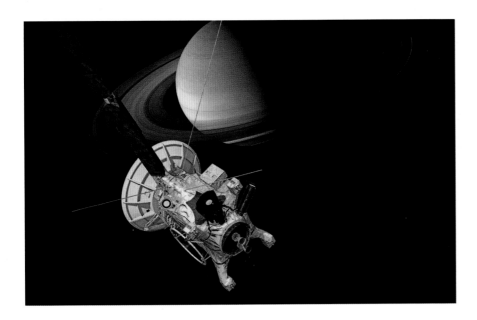

Saturn's spectacular ring system makes it the most recognizable planet in the Solar System. Although Galileo could see something to each side of the most distant planet then known, he was not able to discern what he was seeing. In 1655 Dutch astronomer Christiaan Huygens became the first astronomer to describe them as a disk surrounding Saturn, and Italian-French astronomer Giovanni Cassini was the first to see gaps in the rings; the largest of these gaps was later named the Cassini Division.

The exact nature of the rings remained a mystery for several centuries. In a masterpiece of mathematical reasoning, James Clerk Maxwell argued in 1859 that the rings could not be a solid disk or they would break apart. He proposed that they were instead composed of countless small particles, all independently orbiting the planet; a proposal which proved to be correct.

The Cassini space probe was launched by NASA in 1997 and arrived at Saturn in mid-2004. It has been orbiting Saturn since, giving us our most detailed images of the planet, its ring system, and its many moons.

Shown opposite (top) is an image of Earth taken by Cassini, with Saturn's rings in the foreground. The Earth and Moon form the bright dot at lower right. The other image (bottom) was taken by the MESSENGER spacecraft, which was sent to orbit Mercury. At this closer distance, MESSENGER is able to see the Earth and Moon as two separate objects.

Above: Artist's impression of the Cassini space probe approaching Saturn.

Opposite above: Earth seen from the Cassini space probe at Saturn.

Opposite below: Earth and Moon seen from MESSENGER at the distance of Mercury.

GEYSERS ON ENCELADUS

Cassini Space Probe
Visible Light

- Spectacular plumes of water shoot out into space from the surface of Saturn's moon Enceladus. Discovered in 2005 by the Cassini spacecraft, these geyser-like jets originate from a region near the moon's south pole and vent from cryovolcanoes; volcanoes which erupt volatiles such as water, ammonia or methane instead of molten rock. More than 100 geysers have been found, with water being the main component. Other volatiles are also emitted in the plumes, together with some solid material including sodium chloride and ice particles. It has been estimated that approximately 200 kilogrammes of material is emitted every second, with some of the water falling back onto the surface of Enceladus as snow. However, most of the material erupted escapes the moon, and supplies the majority of the material which makes up Saturn's E ring.

   Enceladus was discovered in 1789 by William Herschel, although little was known about it until the two Voyager spacecraft flew past in the early 1980s. Starting in 2005, the Cassini spacecraft has made multiple fly-bys of Enceladus to study its surface in more detail. It is Saturn's sixth largest moon, with a diameter of about

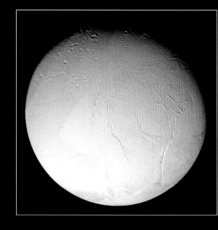

Above: Enceladus imaged by the Cassini spacecraft.

500 kilometres, one tenth the diameter of Saturn's largest moon Titan (see page 41). Due to the water which falls back onto the moon from its plumes as snow, the surface is covered in fresh, clean ice which reflects most of the sunlight falling upon it. Cassini has also discovered that Enceladus has internal heat escaping from it; the source of internal heating is tidal, as with Jupiter's moon Io (see page 36). Very few impact craters are seen in the region near the south pole. This and its internal heat shows that Enceladus is geologically active.

In 2014, NASA announced that Cassini had found evidence for a sub-surface ocean of liquid water near the moon's south pole. It has been calculated that this ocean has a depth of around 10 kilometres, making Enceladus another prime candidate among the bodies in our Solar System which may harbour life.

HYPERION

Cassini Space Probe
Visible Light

- Hyperion, discovered in 1848, is another of Saturn's moons and is noted for its irregular shape, its chaotic rotation and its unusual appearance. It is named after Hyperion, one of the 12 children of Gaia and Uranus in Greek mythology. Hyperion was the first non-round moon to be discovered; it measures about 360 kilometres by 266 kilometres by 205 kilometres. The only larger moon known which also has an irregular shape is Neptune's moon Proteus.

  Voyager 2 photographed Hyperion in 1981, but only from a distance. Early images taken by the Cassini spacecraft suggested that Hyperion had an unusual appearance, but the true extent of its strangeness was not evident until Cassini performed a targeted fly-by in September 2005. Since then Cassini has made three more close approaches; in August 2011, September 2011 and May 2015. The detailed images obtained during these fly-bys have revealed a strange, sponge-like appearance.

  Based on measurements made by Cassini, it would appear that Hyperion is about 40 percent empty space. Its surface is covered in deep craters with sharp edges; at the bottom of each crater is dark material. It has been suggested that its weak surface gravity means that impacting objects tend to compress the surface rather than excavating it, and most of the material blown off the surface never returns. It is thought that this is what is responsible for its sponge-like appearance. It is also known that Hyperion is composed largely of water ice with very little rock. Its chaotic rotation is attributed by some to be due to its close proximity to Titan, Saturn's largest moon.

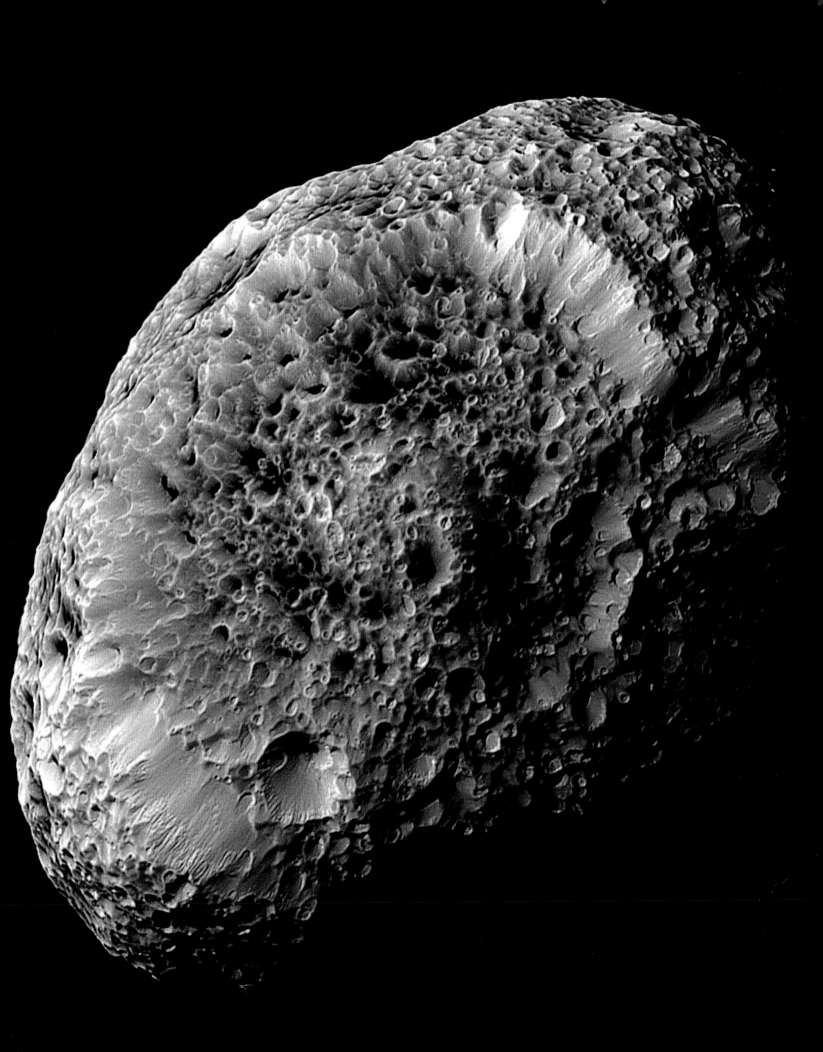

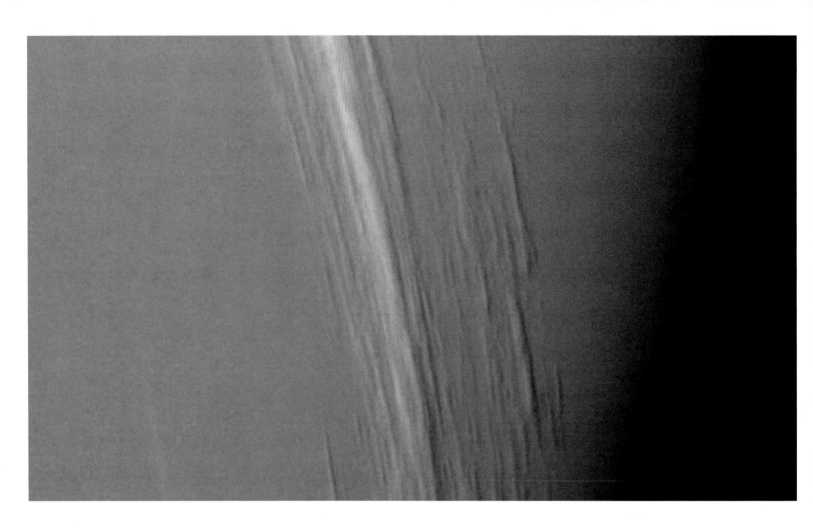
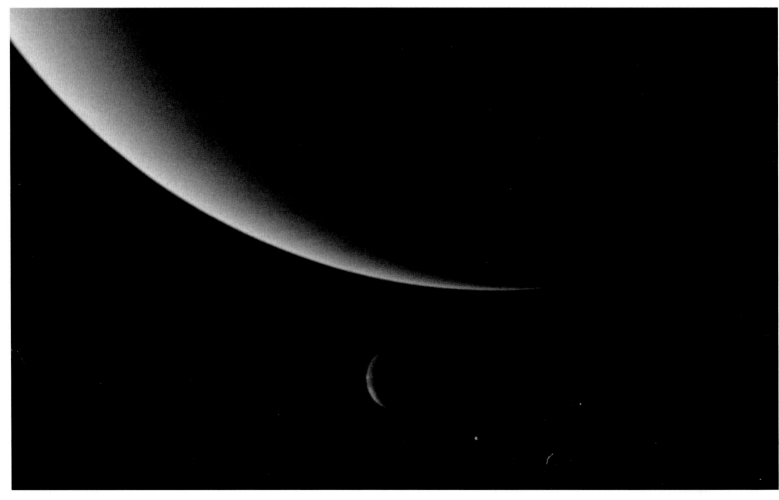

NEPTUNE IMAGED BY VOYAGER 2

Voyager 2
Visible Light

- In August 1989 the Voyager 2 space probe reached its final Solar System target, the planet Neptune. This planet was discovered in 1846 using the power of mathematics and Newton's equation of gravity. After Uranus was discovered in 1781, the following decades allowed astronomers to study its orbit in greater detail. In 1821 Alexis Bouvard published detailed tables of Uranus's orbit based on its distance from the Sun. Subsequent observations, however, revealed that Uranus was not where it should be in the sky; it seemed to be a few degrees away from its predicted position. To put it another way, it was not following its expected orbit.

All the planets in the Solar System orbit the Sun, but their orbits are affected by the gravitational effect of the other planets. Given the position of Uranus, the two planets which should affect its orbit the most are Saturn and the Solar System's largest planet, Jupiter. But, even after the gravitational tugs exerted by Saturn and Jupiter had been taken into account, Uranus was still showing an anomalous orbit. Bouvard suggested that the orbit of Uranus was being affected by an unknown object, another planet.

Two mathematicians set about calculating the position of this hypothesized planet, Englishman John Couch Adams and Frenchman Urbain Le Verrier. They sent their calculations to Cambridge and Berlin observatories respectively, and in September 1846 Berlin Observatory won the race when Johann Galle discovered the new planet, which was named Neptune.

Opposite above: The clouds of Neptune, seen by Voyager 2 two hours before closest approach.

Opposite below: Neptune and Triton, its moon, photographed by Voyager 2 three days after fly-by.

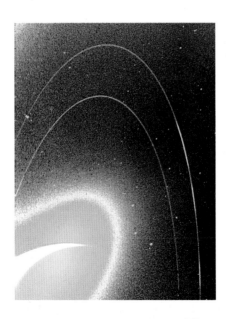

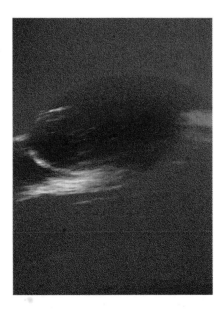

Left: Neptune's rings seen by Voyager 2. The bright glare is due to overexposure from Neptune's reflected light.

Right: When Voyager 2 flew past Neptune in 1989, one of the most surprising discoveries was that it had a Great Dark Spot, comparable in size to Jupiter's Great Red Spot – a rotating storm system. When the Hubble Space Telescope viewed the planet in 1995, the spot had vanished, but a similar spot had appeared in the Northern Hemisphere of the planet.

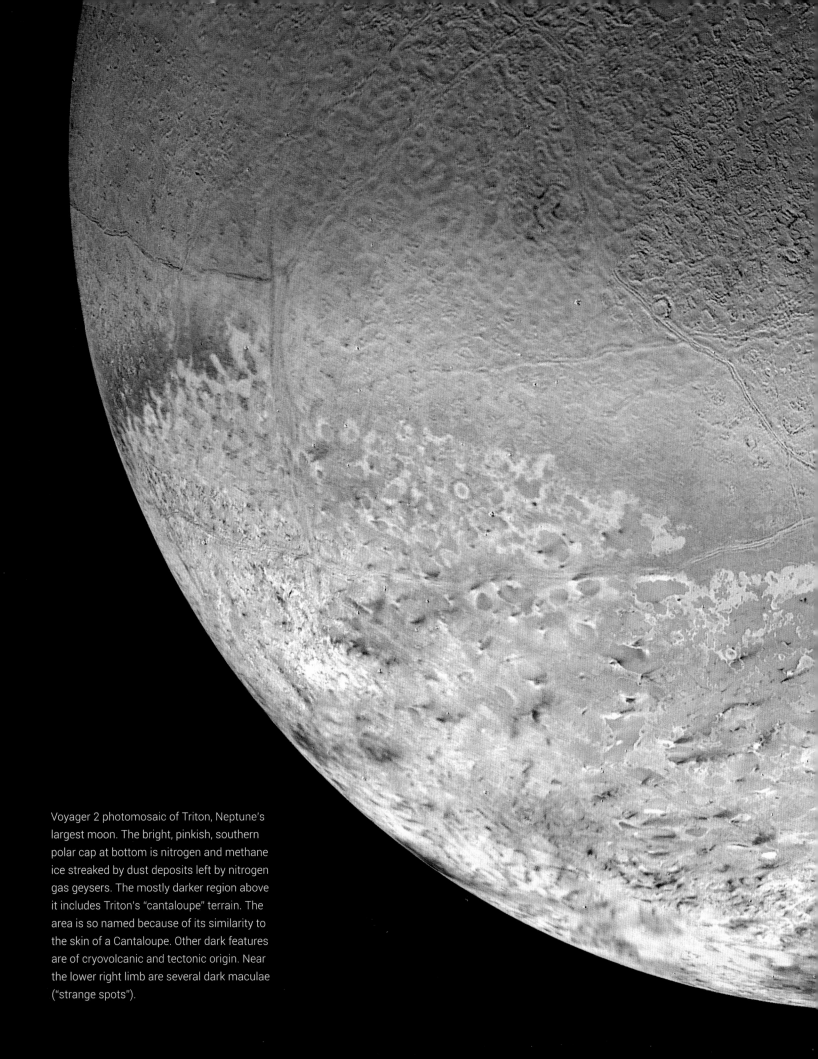

Voyager 2 photomosaic of Triton, Neptune's largest moon. The bright, pinkish, southern polar cap at bottom is nitrogen and methane ice streaked by dust deposits left by nitrogen gas geysers. The mostly darker region above it includes Triton's "cantaloupe" terrain. The area is so named because of its similarity to the skin of a Cantaloupe. Other dark features are of cryovolcanic and tectonic origin. Near the lower right limb are several dark maculae ("strange spots").

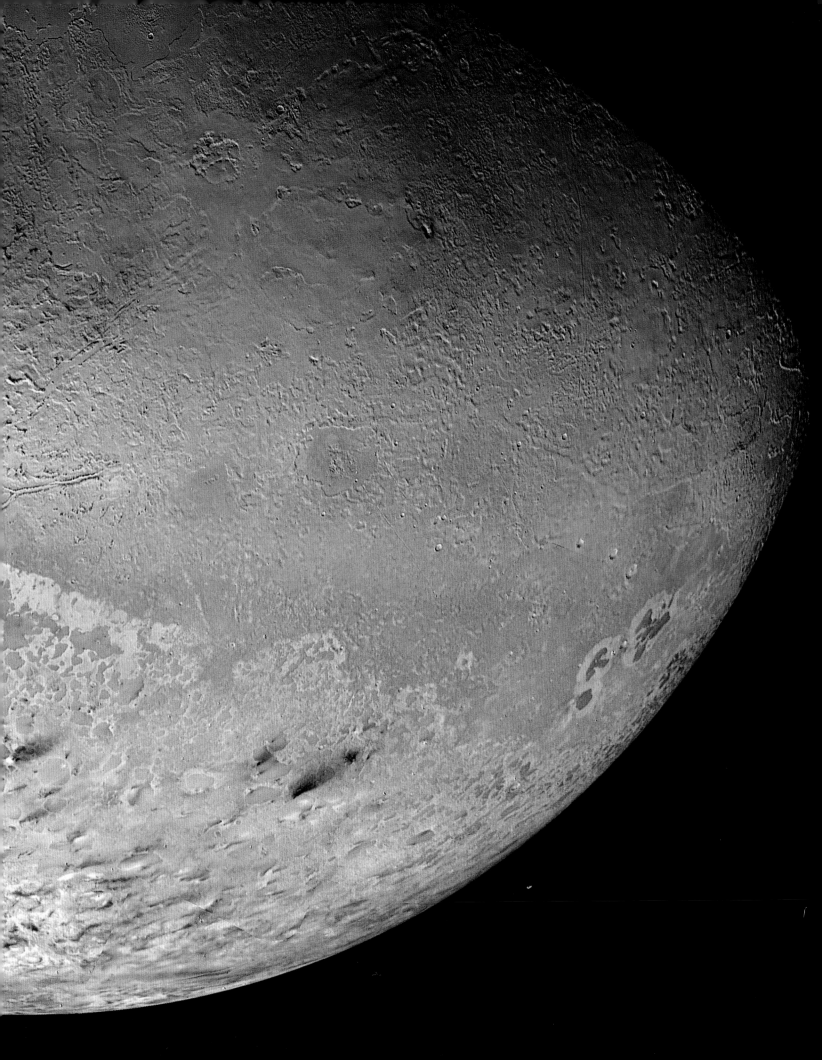

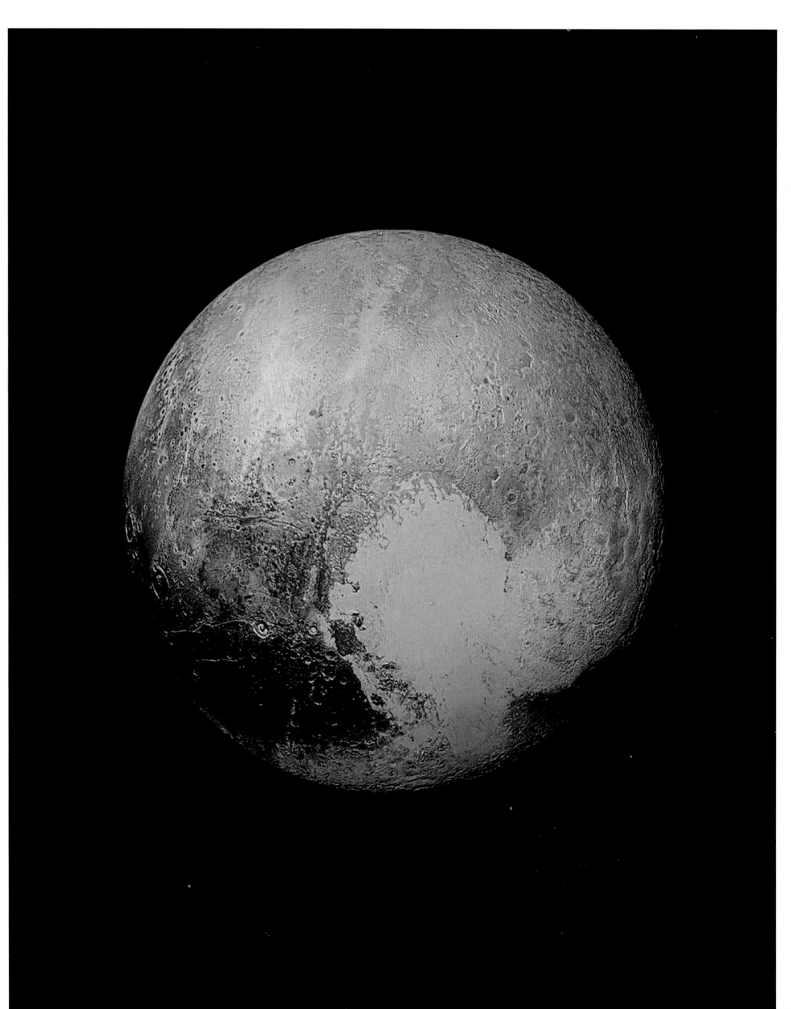

THE DWARF PLANET PLUTO

New Horizons Probe
Visible Light

- The New Horizons space probe left Earth in January 2006 on a nine-year voyage to study Pluto. At the time of New Horizons' launch, Pluto was classified as the ninth and most distant planet in the Solar System. However, this changed in August of the same year, when the International Astronomical Union (IAU) decided to reclassify Pluto as a "dwarf planet". Pluto had been discovered in 1930 by Clyde Tombaugh, working at the Lowell Observatory in Flagstaff, Arizona. By comparing hundreds of pairs of images of the sky, he found a tiny speck of light which had moved between two photographs; the astronomical community agreed that he had discovered a new planet and it was named Pluto.

In the 1950s it was suggested by planetary astronomer Gerard Kuiper that short-period comets came from a reservoir of icy bodies just beyond the orbit of Pluto. This reservoir became known as the Kuiper belt. In the 1990s astronomers started finding several of these Kuiper belt objects, many of them similar in size to Pluto. These discoveries led to a debate about Pluto's status as a planet. In 2005 Eris was discovered, a Kuiper belt object that was found to be more massive than Pluto. The following year, the IAU voted to reclassify Pluto as a minor planet, along with several other Kuiper belt objects and asteroids.

After its nine-year journey, New Horizons flew past Pluto on 14 July 2015 within 12,600 kilometres of its surface. It has obtained the first ever detailed images of the surface of Pluto and its five moons.

Below: In the clean room at Kennedy Space Center's Payload Hazardous Servicing Facility, technicians prepare the New Horizons spacecraft for a media event before its launch.

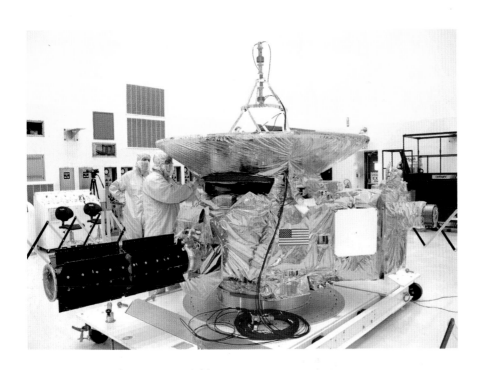

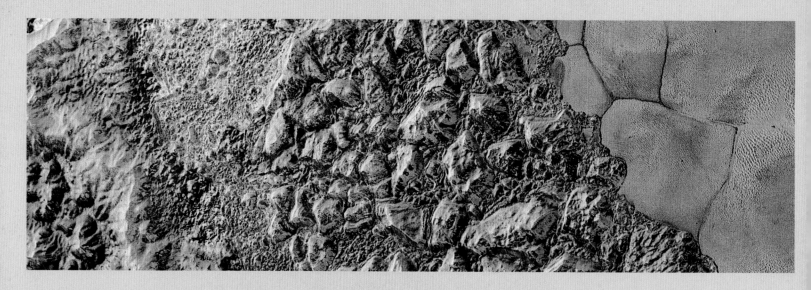

Above: New Horizons captured this high-resolution view of Pluto's surface which shows the al-Idrisi Montes, mountainous highlands thought to be composed primarily of blocks of water ice. On the right are ice plains that make up the heart-shaped feature, known as Sputnik Planum.

Below: A dark, mysterious, north polar region nicknamed Mordor Macula tops this high-resolution portrait of Pluto's largest moon Charon, seen from New Horizons near its closest approach on 14 July 2015.

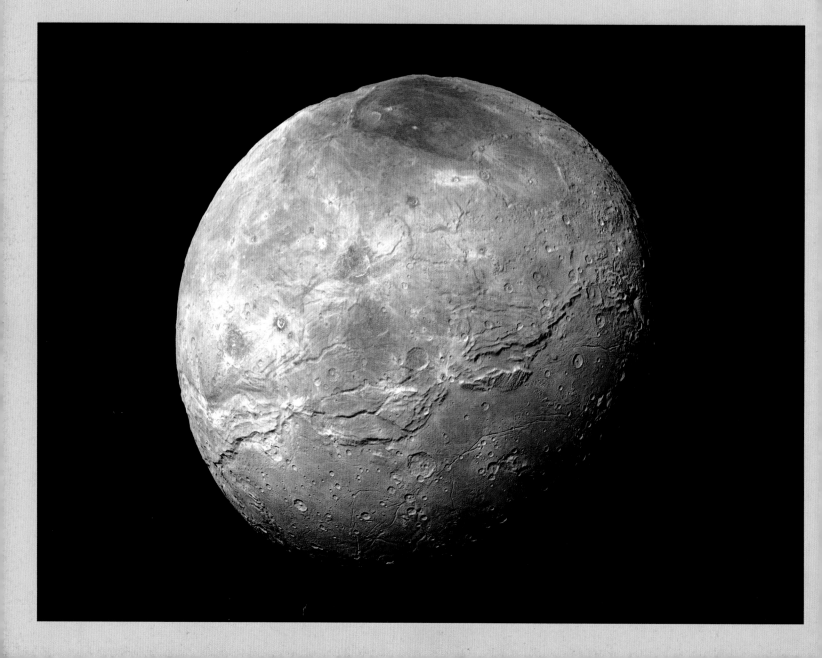

Observations of the approach to Pluto by New Horizons, April–July 2015

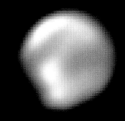
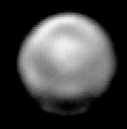

First Pluto-Charon colour image, 14 April 2015.

75 million km from Pluto, 12 May 2015.

50.5 million km from Pluto, 2 June 2015.

Features visible on surface, 15 June 2015.

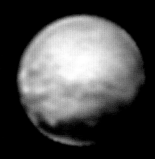
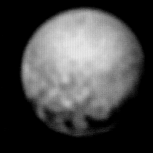
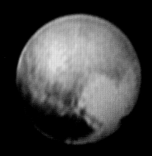
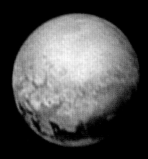

13.5 million km from Pluto, 3 July 2015.

A colour version, 3 July 2015.

A heart on Pluto, 7 July 2015.

Signs of geology, 10 July 2015.

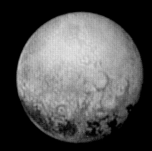
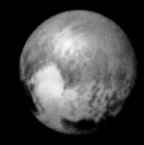
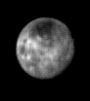

Last photo of Charon-facing hemisphere, 11 July 2015.

Pluto and Charon in false colour, 13 July 2015.

Last image sent before closest approach, 13 July 2015.

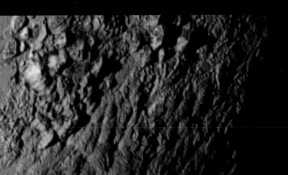

From the base of the heart-shaped feature on Pluto, showing a mountain range with peaks jutting as high as 3,500 metres above the surface of the icy body, 15 July 2015.

COMET LOVEJOY

From the International Space Station
Visible Light

- Comet Lovejoy, more correctly known as C/2011 W3, passed through the Sun's corona on 16 December 2011. The incredible photograph of Comet Lovejoy shown here was taken from the International Space Station by astronaut Dan Burbank in 2011, as the comet approached the Sun.

  Comet Lovejoy is a long-period comet which takes just over 600 years to orbit the Sun. It was discovered by Australian amateur astronomer Terry Lovejoy in late November 2011 using a 20-centimetre telescope and a sensitive digital camera known as a Charge-Coupled Device (CCD). In fact, it was the third comet discovered by Lovejoy, and since then he has discovered two more; so we need to specify C/2011 W3 to ensure that we are referring to this particular comet. C/2011 W3 was independently verified on 1 December; it was the first sun-grazing comet to be discovered by ground-based telescopes in more than 40 years.

  It is estimated that C/2011 W3 lost a significant fraction of its nucleus as it grazed the Sun. Before passing through the Sun's corona it is believed that the comet's nucleus measured some 500 metres across; afterwards this had shrunk to between 100 and 200 metres. A significant outburst of dust was observed on 19 December, just three days after its passage through the corona. Although it appeared to survive its passage, some astronomers believe that the comet subsequently broke up entirely and has completely disappeared. If not, it is calculated that it will return around the year 2633.

Opposite: Commander Dan Burbank captured this spectacular photograph of Comet Lovejoy from about 380 kilometres above the Earth's horizon, onboard the International Space Station.

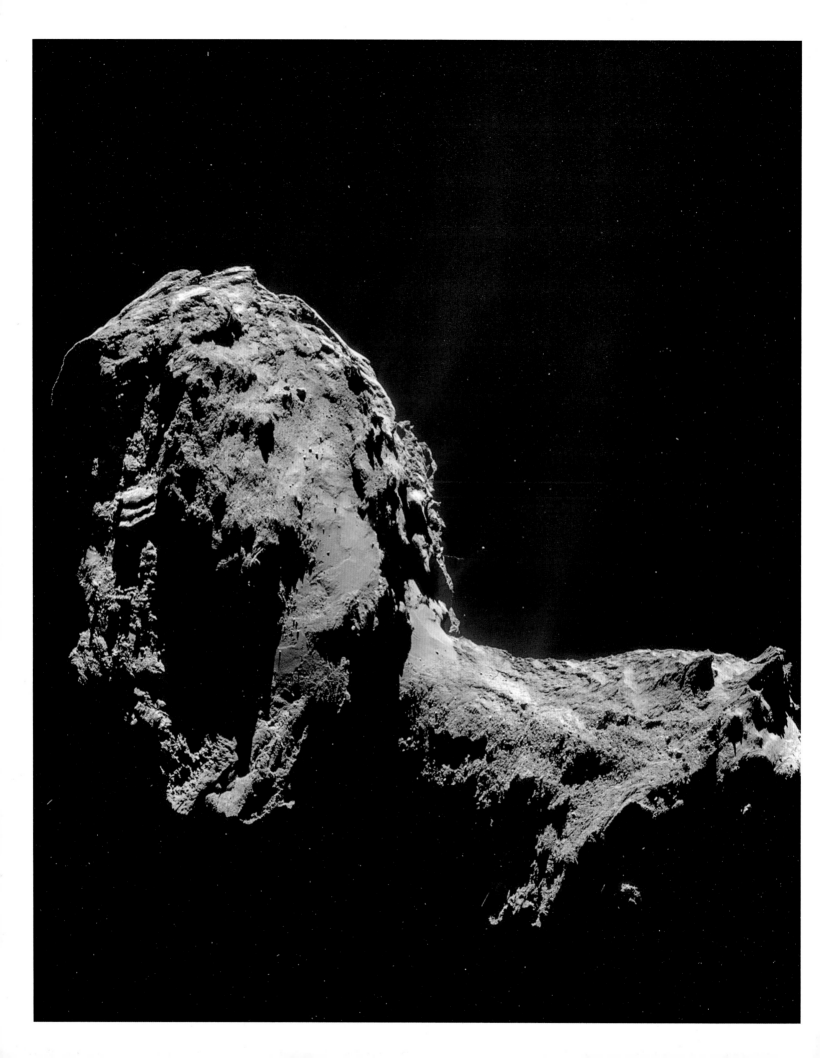

COMET 67P AND THE ROSETTA MISSION

Rosetta Mission
Visible Light

- In March 2004 the European Space Agency (ESA) launched a space probe with the audacious target of landing on a comet. More than 10 years later, in November 2014, the Philae lander successfully touched down on Comet 67P, capping a remarkably successful mission which achieved a number of firsts in space exploration. Comet 67P's full name is comet 67P/Churyumov-Gerasimenko, named after the two Soviet astronomers who discovered it in 1969. 67P is an example of a short-period comet. It originates from the Kuiper belt and its current period is just under 6.5 years.

  Rosetta flew past Mars in February 2007 to perform a gravity assist, using Mars' momentum to speed itself up. After two fly-bys of asteroids in September 2008 and July 2010 the spacecraft was put into hibernation to conserve battery power as its solar panels were too far from the Sun to continue powering it. After a 31-month hibernation it was successfully woken up in January 2014 as it approached 67P.

  Rosetta arrived at the comet in August 2014, going into orbit about it approximately 30 kilometres from its nucleus and taking high resolution images such as the one shown opposite. Rosetta continued in orbit around 67P as the comet reached its closest position to the Sun in August 2015. One of Rosetta's most important discoveries is that the comet's water composition is substantially different from Earth's, making it unlikely that comets with 67P's composition brought water to our planet.

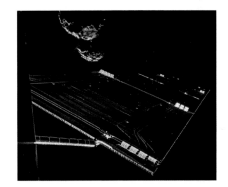

Above: Rosetta took this selfie 16 kilometres from the surface of comet 67P. At the top of the frame, dust and gas stream away from the comet's double-lobed nucleus and sunlight reflects off one of Rosetta's 14-metre long solar arrays.

Below: ESA's Rosetta spacecraft, which reached comet 67P in August 2014.

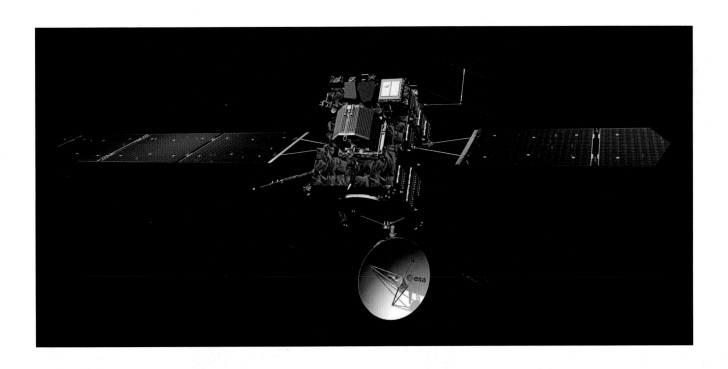

THE PHILAE LANDER

Rosetta Mission and Philae Lander
Visible Light

On 12 November 2014 the Philae lander became the first space probe to successfully land on the nucleus of a comet. After a remarkable 10-year journey, ESA's Rosetta mission dropped the tiny probe towards its target, comet 67P. The image shown at right was taken by Rosetta as Philae descended some 25 kilometres through the blackness of space towards its landing site. However, due to the comet's extremely weak gravity, Philae bounced after first touching the surface. It rose to an altitude of about 1 kilometre, rebounding a second time before finally coming to a halt.

Unfortunately, due to these two bounces, Philae's final landing place was not optimal. Although not damaged, it sits in the shadow of a nearby cliff or crater wall leading to its on-board solar panels not receiving sufficient sunlight to power the lander's secondary rechargeable battery. The primary batteries were designed to last about 60 hours, and contact was lost on 15 November. As 67P approached the Sun the landing site received more sunlight, and contact was re-established on 19 June 2015. After a few weeks of intermittent contact, it sadly ceased again on 9 July.

During the brief period after Philae's landing it was able to perform a number of its key experiments. Sixteen organic compounds were detected on 67P's surface, four of which were seen for the first time on a comet. Although it had encountered problems, the Rosetta mission and the Philae lander achieved their primary objectives and have greatly increased our understanding of comets.

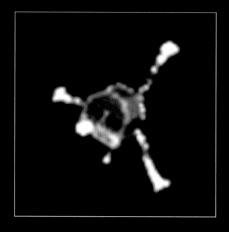

Opposite: A spectacular jet captured by Rosetta's narrow angle camera on 12 August 2015.

Below: The Philae Lander touched down adjacent to a cliff. One of its feet is visible in the foreground.

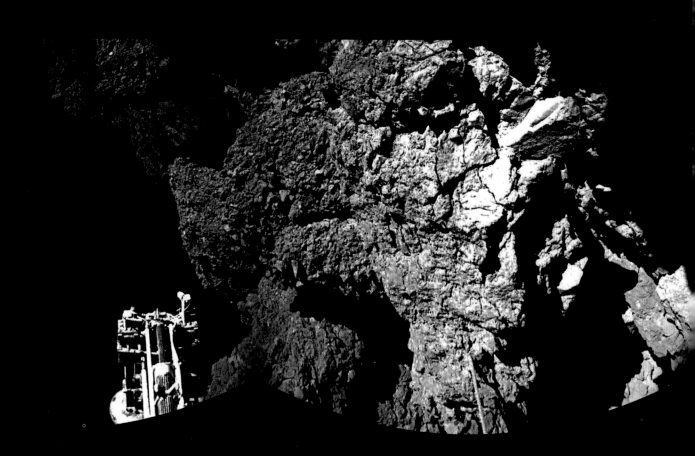

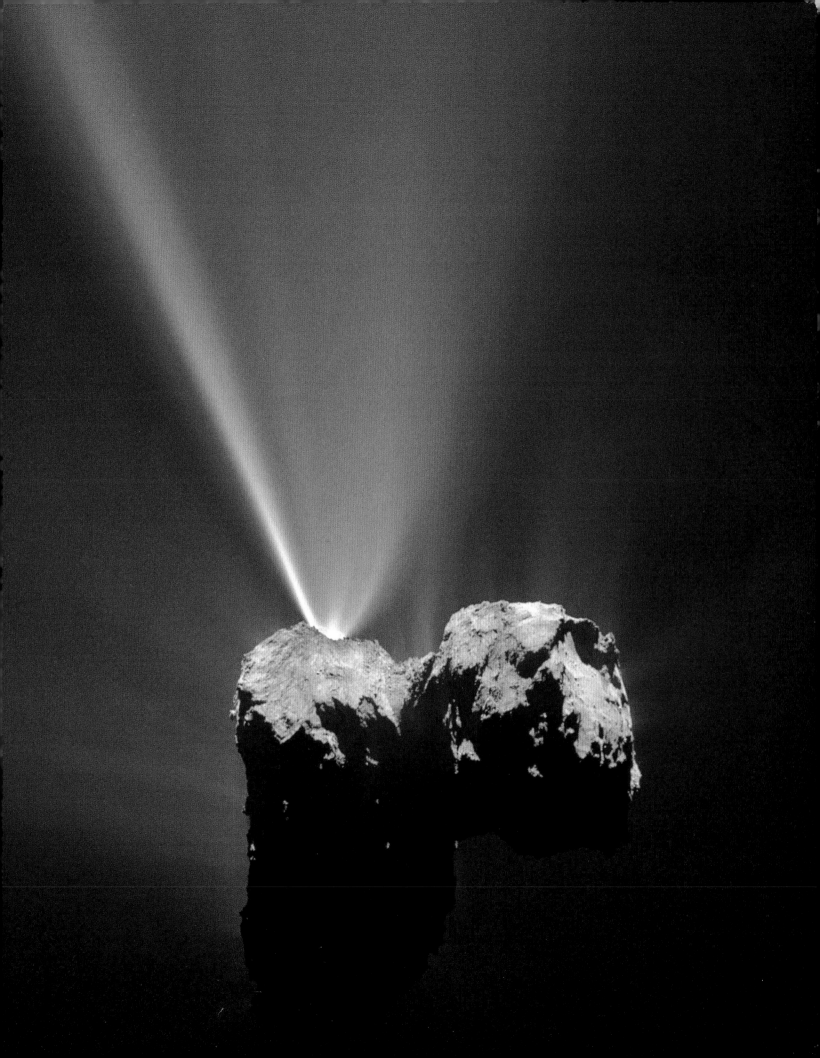

PART 2

EXPLORING THE MILKY WAY

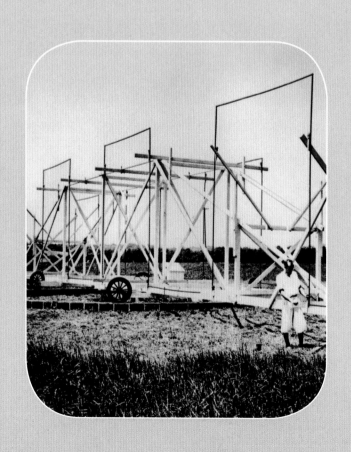

The faint band of light which stretches across our night-time sky has captivated humans for thousands of years. To the naked eye the Milky Way appears as just that, a path of milky whiteness meandering through the stars. It was Galileo who first saw the true nature of the Milky Way. Using a telescope he observed that the band of light was, in fact, thousands of stars too far away to be seen as individual stars by the naked eye. The Milky Way is a vast system of stars, but its true extent and our Sun's place in it remained a mystery until the twentieth century.

One of the reasons our Milky Way's mysteries were so difficult to unravel can be seen in the visible light images shown in this section. The space between stars is not empty, but instead is filled with gas and dust. The dust is effective at absorbing starlight, and this limits how far we can see in any given direction. This absorption of starlight by dust leads to our seeing essentially the same number of stars in the band of the Milky Way no matter which direction we look, and this had led many astronomers to mistakenly conclude that we must be at its centre. It was argued by others that the asymmetrical distribution of globular clusters in the sky showed that we were not at the centre; rather we are out in the disk.

For centuries astronomers were only able to explore the Milky Way using visible light, the tiny part of the electromagnetic spectrum to which our eyes are sensitive. Then, in the 1930s, Karl Jansky accidentally discovered radio emission coming from the Sagittarius constellation. By the 1950s radio astronomy had been born, and since then every part of the spectrum has been opened up to astronomy, from high energy gamma rays to low energy radio waves. In order to see gamma rays, X-rays and ultraviolet light from beyond our Earth we need to go into space, as these wavelengths cannot penetrate the ozone layer in our atmosphere. Many of the X-ray images shown here were taken by NASA's most recent X-ray observatory, the Chandra X-ray Observatory. These short wavelengths allow us to see some of the most energetic processes in our Galaxy, such as exploding stars and the formation of black holes.

Infrared and microwaves are also absorbed by our atmosphere, in this case by water vapour. Many of the infrared images shown here were obtained by NASA's Spitzer Space Telescope and ESA's Herschel Space Observatory. Without these space-based infrared telescopes, it would be impossible to penetrate through the thick clouds of gas and dust that enshroud sites of star formation. Infrared light has also enabled us to see our Milky Way as it would appear without the effects of dust; such images clearly show a disk of stars with a central bulge, with our Sun about a third of the way out in the disk.

As the images in this section show, our understanding of our home Galaxy has progressed beyond measure by our ability to explore it at all of these different wavelengths. Whether it is X-rays captured by Chandra, visible light captured by the Hubble Space Telescope, or infrared captured by Spitzer, a multi-wavelength exploration of our Milky Way has truly enabled us to understand the huge variety of processes which occur, from stellar birth to stellar death.

Opposite: Karl Jansky with his "merry-go-round" antenna, the very first radio telescope. It detected the hiss from our Galaxy as a series of peaks on a chart as the antenna spun round.

its glory and understand our place in it.

The twentieth century has seen us open up the entire electromagnetic spectrum to observation. We now know that visible light, the part to which our eyes are sensitive, is just a tiny part of a much larger canvas which ranges from radio waves at the lowest energies to gamma rays at the highest energies. X-rays, ultraviolet light, infrared and microwaves fall between these two extremes. Each part of this spectrum can tell us different things about our Milky Way, as different physical processes are responsible for emitting at the different energies.

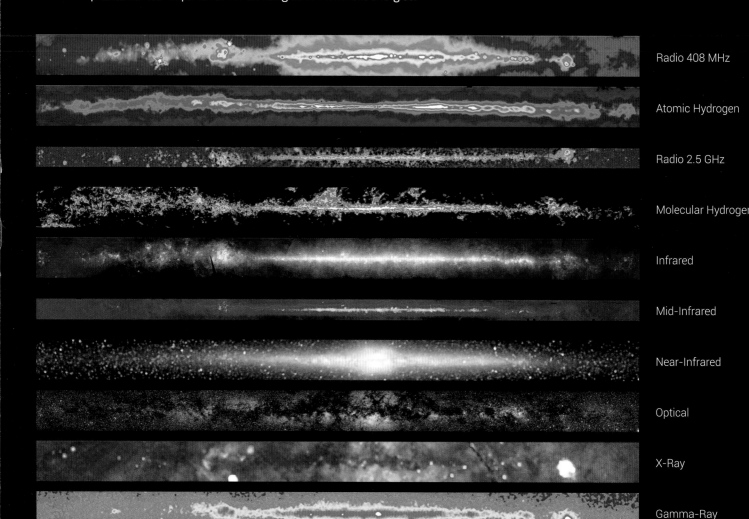

Radio 408 MHz

Atomic Hydrogen

Radio 2.5 GHz

Molecular Hydrogen

Infrared

Mid-Infrared

Near-Infrared

Optical

X-Ray

Gamma-Ray

At radio wavelengths we can see emission from neutral atomic hydrogen gas, molecular hydrogen gas, as well as emission from fast-moving electrons accelerated by supernovae explosions. In the infrared we can see emission from dust and from cool stars which are otherwise hidden from view by the dust, and at X-rays we see emission from hot gas at millions of degrees. At the highest energies we see emission from cosmic ray collisions, as well as merging neutron stars and black holes.

Below: 2MASS (2 Micron All-Sky Survey) view of the centre of the Milky Way, using observations from Mount Hopkins Observatory, Arizona in the Northern Hemisphere, and Cerro Tololo Inter-American Observatory, Chile, in the Southern Hemisphere.

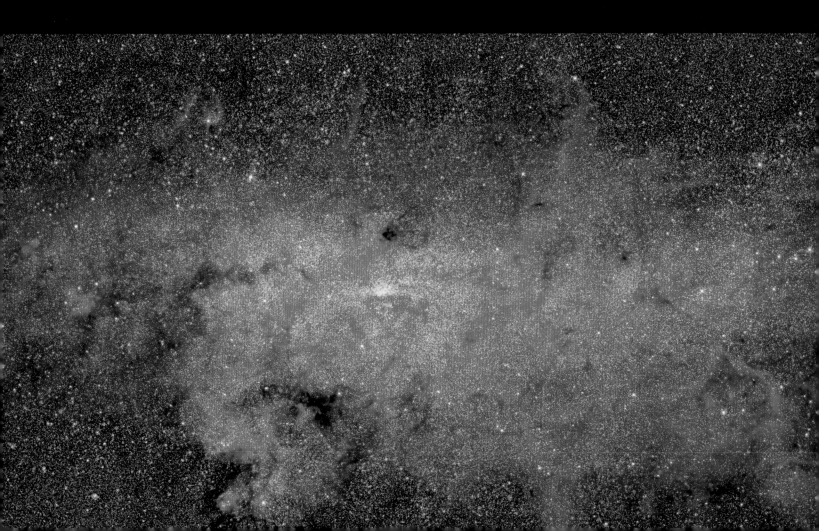

THE MILKY WAY FROM SPITZER

Spitzer Space Telescope
Infrared

- The most detailed infrared image of the plane of our Milky Way Galaxy has been pieced together from more than 800,000 images taken by various cameras on the Spitzer Space Telescope. The image shown below is just the central part of that huge image, it spans eight degrees (roughly 16 times the diameter of the full Moon), with the Galactic centre in the middle of the image. The full mosaic of the entire Galactic plane (shown at right) is 15 times wider, 120 degrees wide. Much of what you see in these images is hidden from our view in visible light due to the gas and dust in our Galaxy.

  These false-colour images highlight three separate components in the Galactic plane. The many blue specs that you can see are individual stars in our Milky Way. Many of these would not be seen in visible light; either because they would be hidden by gas and dust, or they are too cool to be emitting at visible light wavelengths. Shown in red is emission from large graphite dust particles; these are associated with regions of star formation and also mixed with the gas between the stars.

  Shown in green is emission from a different type of dust – tiny dust grains which are known as polycyclic aromatic hydrocarbons (PAHs). These carbonate dust particles are actually quite common on Earth, they are the sooty product of incomplete combustion. In space we find that they are spread throughout the Galaxy, even rising above the mid-plane in long filaments.

Right: The full mosaic of the Milky Way. Green filaments of light are found in star-forming regions, red patches indicate dust, blue specks are individual stars.

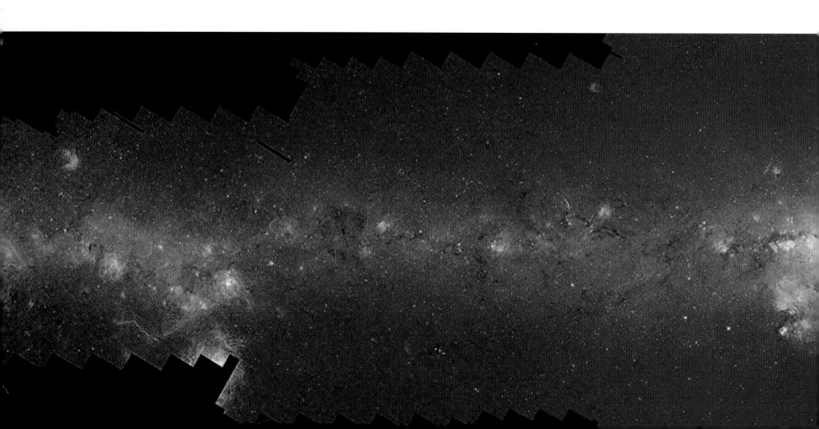

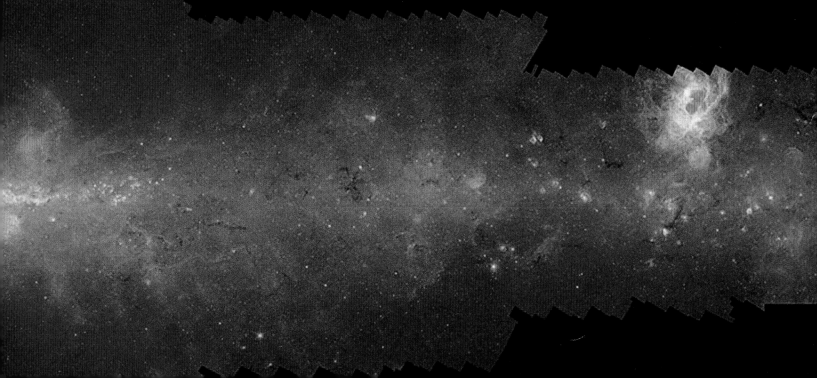

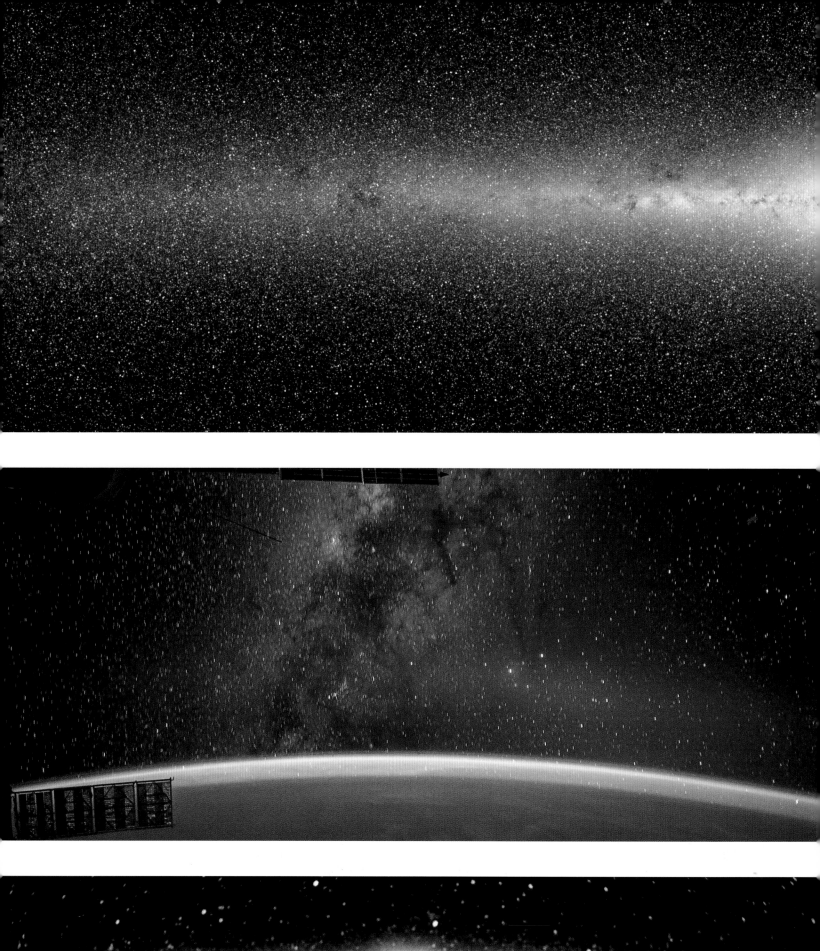
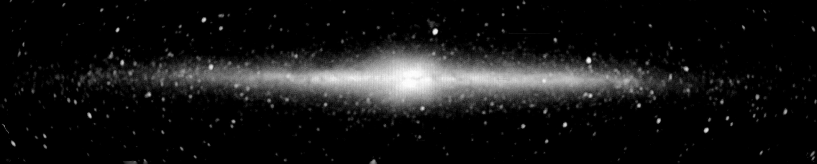

THE MILKY WAY FROM COBE AND 2MASS

Cosmic Background Explorer and Two Micron All-Sky Survey
Infrared and Millimetre Waves

- For centuries the bulge and disk of our home Galaxy had remained hidden from view. At visible light wavelengths we cannot see very far into the disk of the Milky Way; the same clouds of gas and dust which are the nurseries of new stars also veil from view the true structure of our Galaxy. This veil was lifted in the early 1990s by an infrared camera on board the COsmic Background Explorer (COBE), giving us for the first time this stunning image, shown at bottom left.

  The camera which took the image was called DIRBE – the Diffuse InfraRed Background Experiment. The main purpose of DIRBE was to survey the diffuse infrared sky, the emission from everywhere in the sky at wavelengths from 1.25 microns to 240 microns. By comparison, visible light ranges from about 0.4 to 0.7 microns. By observing the sky in the infrared, DIRBE was able to provide this image of the Milky Way. In the same way that firefighters use infrared cameras to see through the smoke which can fill buildings, astronomers can use infrared light to see through the dust that obscures much of our Galaxy from view.

  The COBE-DIRBE image combines data taken at 1.25, 2.2 and 3.5 microns, shown as blue, green and red. Most of the emission is from cool stars, these show up as white in this three-band image. However, the dust is also evident, it shows up as red across the central parts of the disk, showing that it causes some obscuration, even at these longer wavelengths.

Above: Panoramic view of the night sky as seen by the Two Micron All Sky-Survey (2MASS) which represents three infrared wavelengths: 1.2 microns (blue), 1.6 microns (green) and 2.2 microns (red). The fuzzy patch at lower right is the Large Magellanic Cloud.

Opposite middle: NASA astronaut Reid Wiseman captured this image from the International Space Station and posted it to social media on 28 September 2014, writing, "The Milky Way steals the show from Sahara sands that make the Earth glow orange".

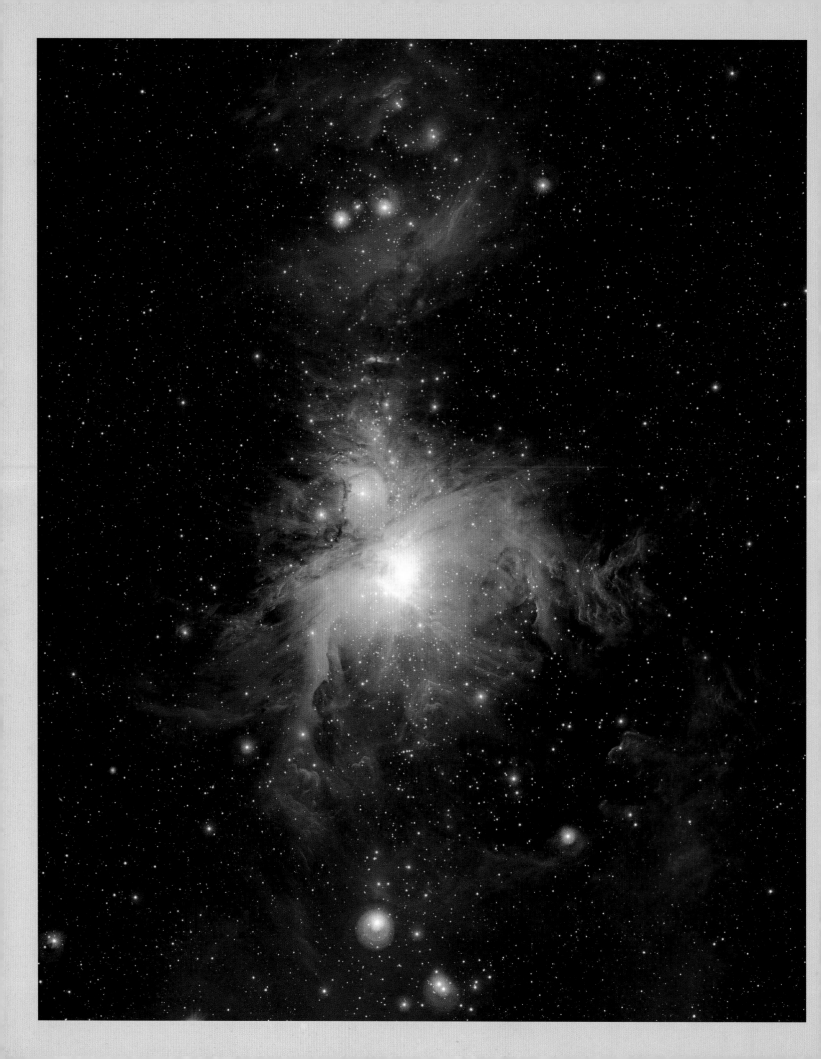

THE ORION NEBULA

Visible and Infrared Survey Telescope for Astronomy
Infrared and Visible Light

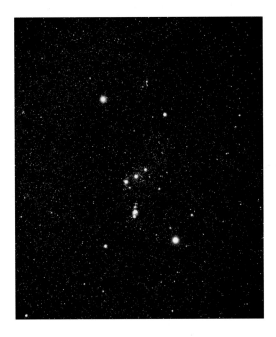

• The Orion Nebula, also known as Messier 42, is the nearest region of massive star formation to us, located some 1,350 light-years away. The nebula is sufficiently close and bright that it is actually visible to the naked eye. The middle "star" of the sword of Orion the Hunter is not, in fact, a star. Instead, you are seeing the glow of ionized gas; this is the Orion Nebula. Because of its proximity to us, it is one of the most studied objects in the sky, and has been imaged across the entire electromagnetic spectrum from radio waves to gamma rays.

The gas which causes the nebula's emission is being ionized by four hot, massive young stars known as the Trapezium (seen in the centre of the image overleaf). Because of their high temperature, these stars are giving off copious amounts of ultraviolet light. This energetic light ionizes the surrounding hydrogen gas causing it to glow, just as ionized gas in a fluorescent light glows. The characteristic red and green glow, seen in visible light overleaf, is due to particular transitions in the electron energy levels of hydrogen gas.

This wide-field view was taken with the VISTA infrared survey telescope at ESO's Paranal Observatory in Chile. This telescope can see the whole nebula and its surroundings and its infrared vision means that it can peer into the hidden dusty regions and reveal the active young stars deep inside. It is immediately apparent how many more stars are visible in the infrared image (at left). This is partly because we are able to see through the gas and dust in the nebula more easily in the infrared, but also because some of the stars are too cool to emit visible light.

Above: Akira Fujii's photograph of Orion, as it appears in an amateur terrestrial telescopic.

Following pages: The Orion Nebula seen by the Hubble Space Telescope in visible light.

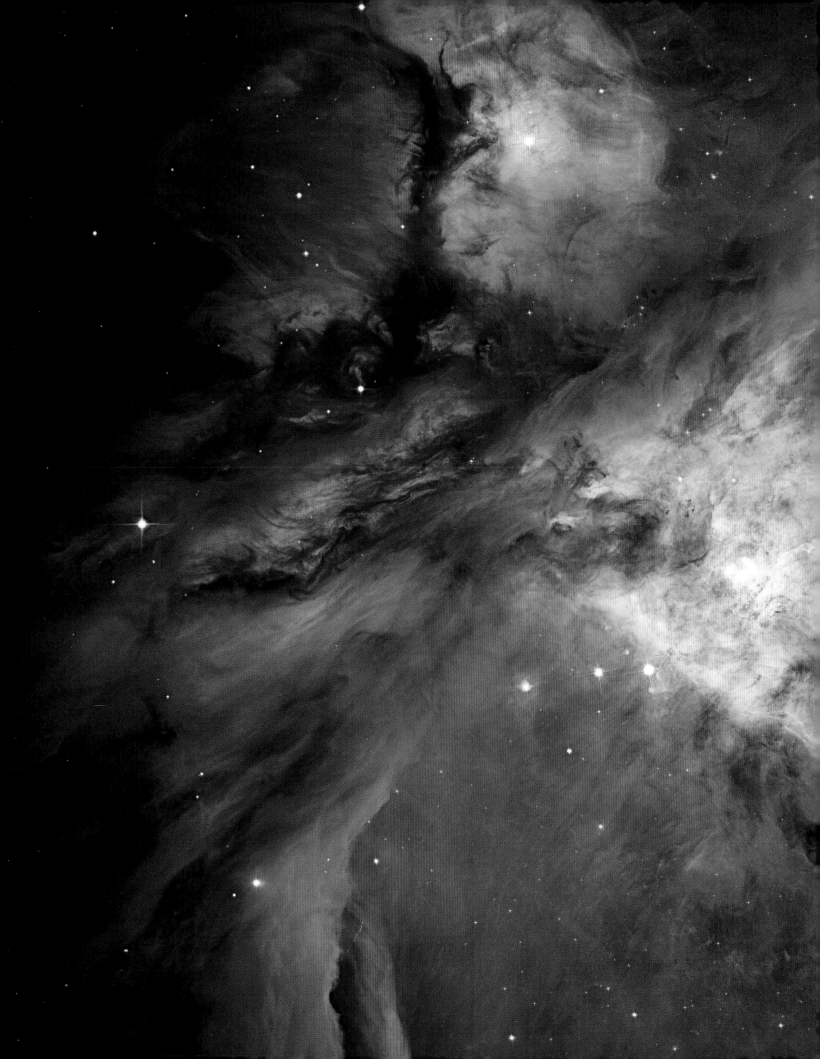

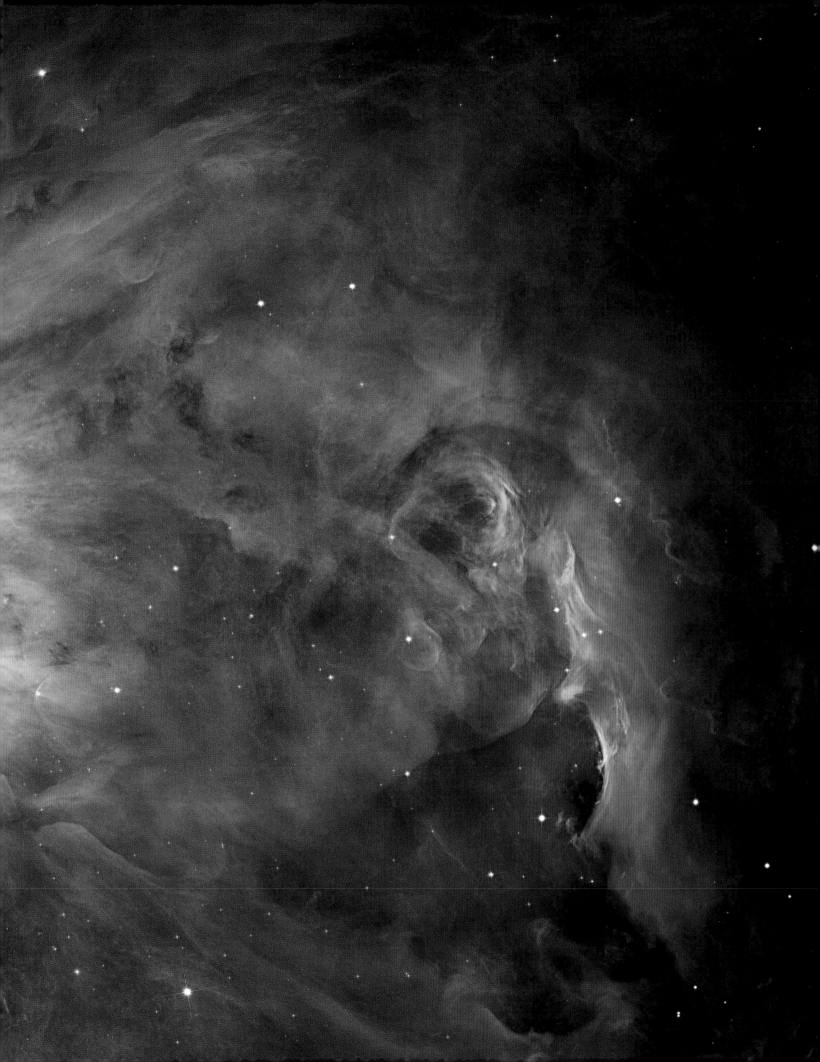

THE TOUCAN'S DIAMOND

Visible and Infrared Survey Telescope for Astronomy
Infrared and Visible Light

● Shining like a brilliant diamond in the southern constellation Tucana is this magnificent object, 47 Tucanae. It is one of the brightest examples of a globular cluster, vast collections of tens of thousands of stars which swarm around each other like bees in a hive. 47 Tucanae is sufficiently bright that it is visible to the naked eye, it was discovered in 1751 by French astronomer Nicholas Louis de Lacaille. It is about 16,000 light-years away and is about 120 light-years across, with a mass of about a million times the mass of our Sun. This colour image of 47 Tucanae was taken by ESO's VISTA (Visible and Infrared Survey Telescope for Astronomy) from the Paranal Observatory in Chile and was released in 2013.

Globular clusters have fascinated astronomers for centuries. American astronomer Harlow Shapley noticed in the early twentieth century that they were not evenly distributed in the sky. Rather, most lie in the part of the sky which we see when we look towards the constellation Sagittarius and neighbouring constellations. At the time, many astronomers believed that the Sun lay at the centre of the Milky Way, but Shapley used the distribution of globular clusters to argue that we are not at the centre, but instead out in the disk of our Galaxy.

Detailed studies of the stars in globular clusters show that they are all of the same age and do not contain any new stars. Globular clusters orbit the bulge of our Milky Way and are amongst the oldest objects in our Galaxy. Studying globular clusters can give us important clues as to how our Galaxy formed and evolved into its present state.

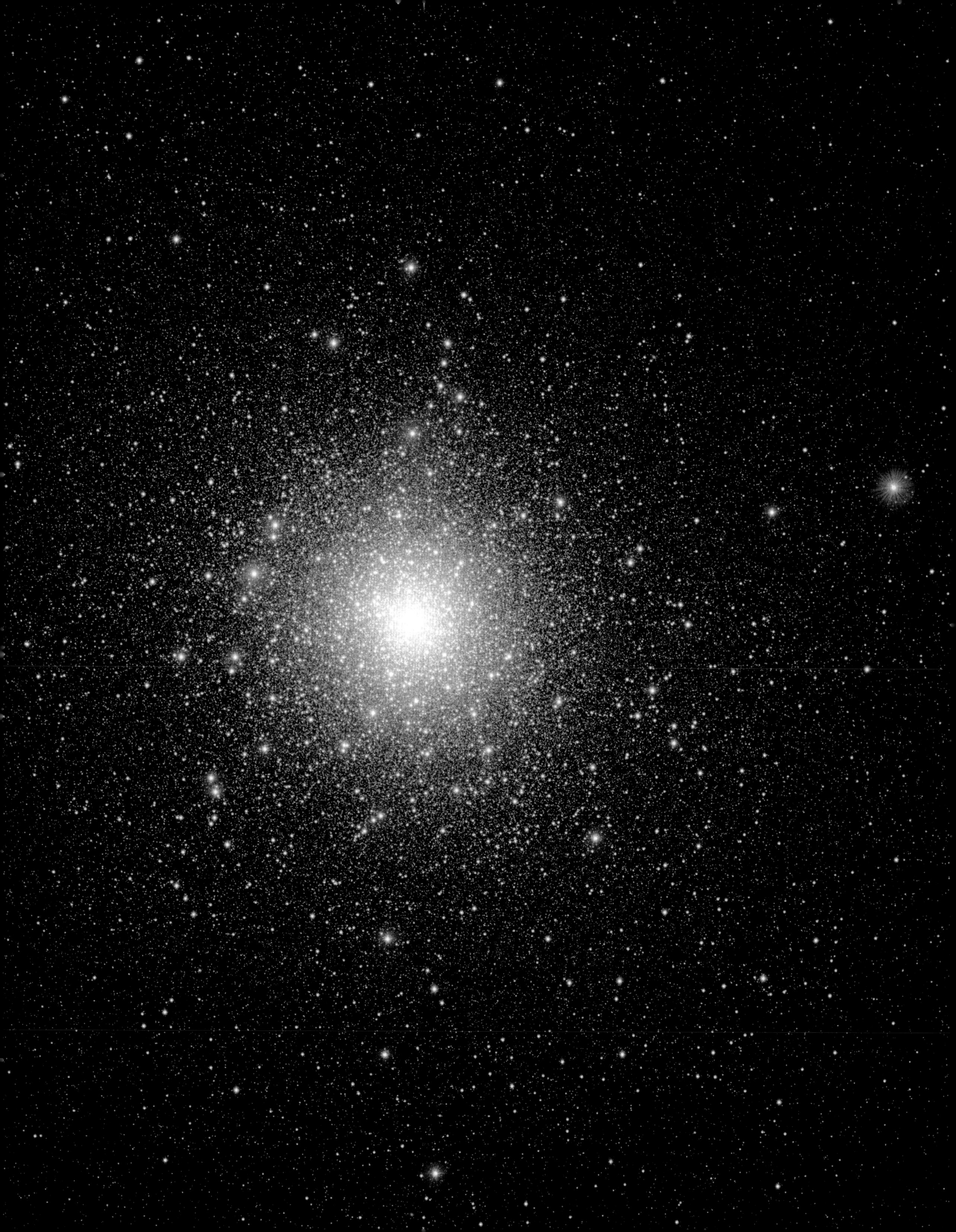

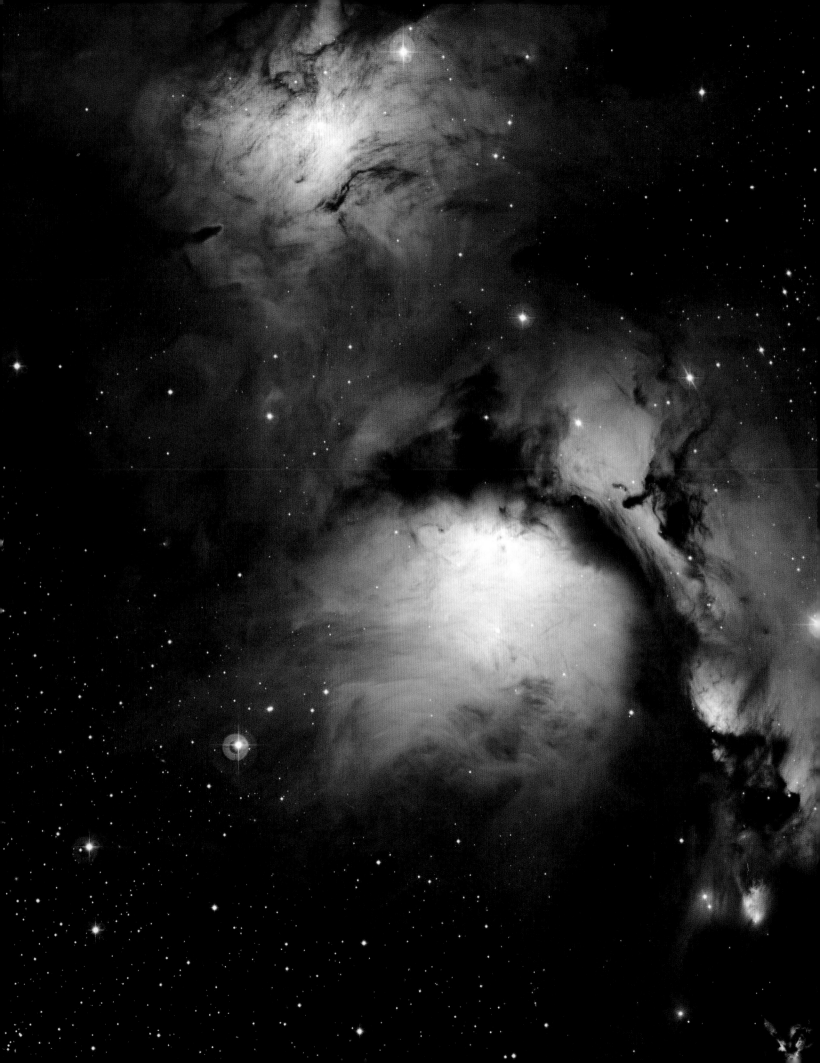

M78 REFLECTION NEBULA

Max Planck Gezellschaft Telescope
Visible Light

- This beautiful reflection nebula is known as Messier 78, or NGC 2068. It is found in the constellation Orion and is the brightest of a group of nebulae that include NGC 2064, NGC 2067 and NGC 2017. The group are all part of the Orion molecular cloud complex, located about 1,600 light-years from Earth. Messier 78 was discovered in 1780 by Pierre Méchain and is one of the brightest and most easily visible reflection nebulae in the sky. Based on its distance and extent in the sky we can calculate that it is about five light-years in size.

  The nebulosity we see in this visible light image is caused by the light from two stars being reflected (or, more correctly, scattered) off dust grains that are contained in a cloud of gas and dust which surrounds the stars. The parts of the cloud which lie to the side of the stars as viewed from Earth are the parts that reflect the starlight, hence the nebulosity appears to the side of the stars even though the cloud actually surrounds the stars. Because blue light is reflected more than green or red light, the reflection nebula appears much bluer than the stars which illuminate it.

  This striking image was captured using the Wide Field Imager camera mounted on the Max Planck Gesellschaft 2.2-metre Telescope at La Silla Observatory in Chile. The real colours were produced by combining many exposures taken separately through red, green and blue filters. Added to this are exposures taken through a narrow filter which isolates emission from ionized hydrogen gas.

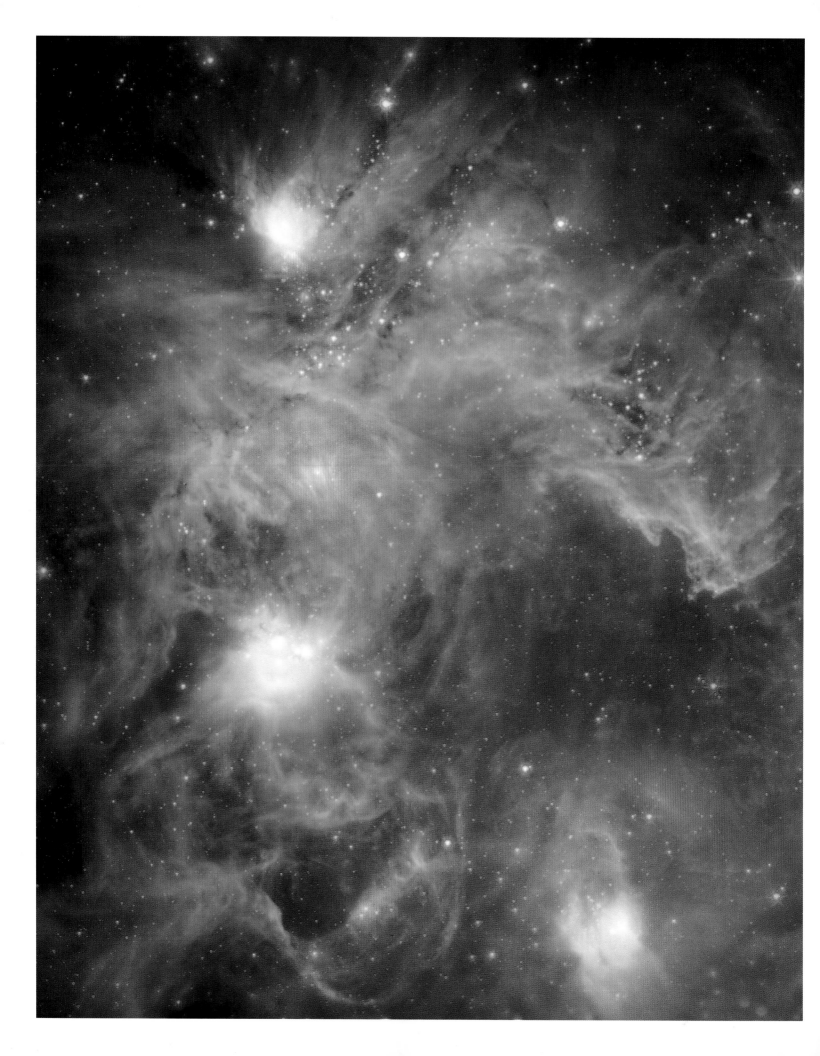

THE MONKEY HEAD

Spitzer and Hubble Space Telescopes
Infrared and Visible Light

● The Monkey Head, more correctly known as NGC 2174, gets its nickname from the resemblance of some of the clouds in this star-forming region to the face of a monkey. However, these features, which can be seen in the visible light image (below), cannot be seen at all in the main image taken in infrared light by NASA's Spitzer Space Telescope. At the longer infrared wavelengths, the clouds of gas and dust become more transparent, allowing us to peer into them to see the sites of recent star formation.

NGC 2174 lies in the northern part of the constellation Orion, and is located about 6,400 light-years from us. Hot, young stars are emitting energetic ultraviolet light and stellar winds (charged particles), and these are carving out clearings in the clouds of gas and dust from which they formed. Whereas these clearings remain hidden from view at visible wavelengths, in the infrared we are able to see them and the hot, young stars responsible.

The reddish spots of light show the next generation of star formation; these are infant stars still swaddled in their blanket of gas and dust. The green is due to emission from organic dust, and blue is from the very hottest areas of dust. This Spitzer image is a combination of three separate infrared wavelengths. Light with a wavelength of 3.5 microns is shown in blue, 8 microns is shown in green, and 24 microns in red (for comparison, visible light ranges from 0.4 to 0.7 microns).

Below: The Monkey Head (NGC 2174) in visible light as seen by the Hubble Space Telescope.

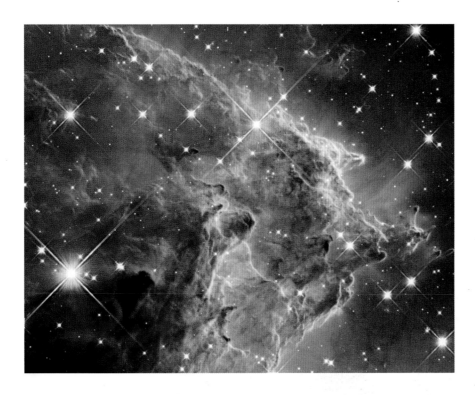

Herschel Space Observatory and Hubble Space Telescope
Infrared and visible light

- The Horsehead Nebula is one of the most imaged features in the sky. The nebula is located just to the south of the star Alnitak, the star farthest to the east on Orion's belt. The main image here shows a composite far-infrared and visible light view from ESA's Herschel Space Observatory and the Hubble Space Telescope. The horsehead can be seen at the far right of the image as a small pink silhouette.

  The nebula gets its name from its obvious resemblance to a horse's head, and was first noted by Scottish astronomer Williamina Fleming in 1888 from a photographic plate taken at the Harvard College Observatory. It is an example of a dark nebula, a thick cloud of dust and gas which is extending up from the main cloud, and blocking the light from stars and glowing gas which lie behind it. The thickness of this filament is illustrated in the near-infrared image at right. Even at

Right: The Horsehead Nebula as viewed at near-infrared wavelengths with the Wide Field Camera 3 on the Hubble. This thick pillar of gas and dust is sculpted by powerful stellar winds blowing from clusters of massive stars located beyond the field of this image. The bright source at the top left edge of the nebula is a young star whose radiation is eroding the surrounding interstellar material.

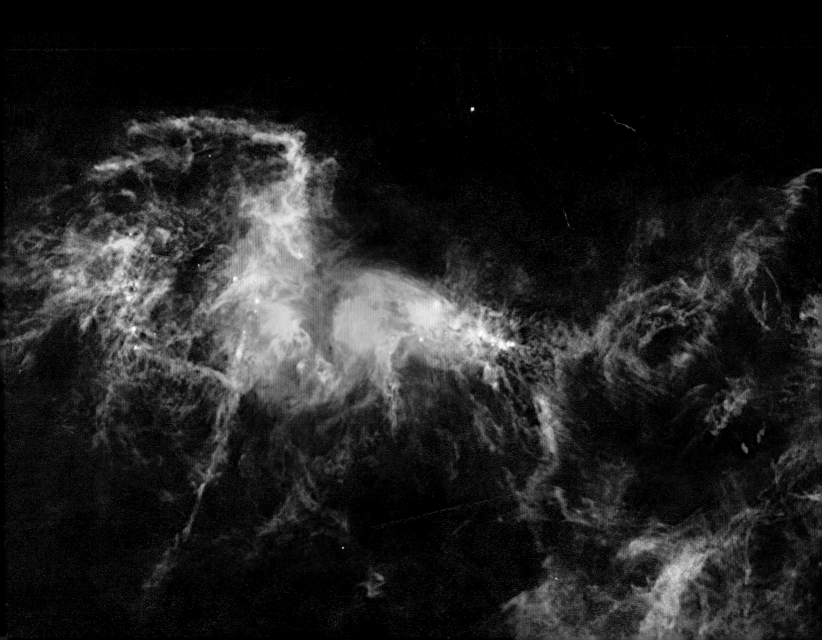

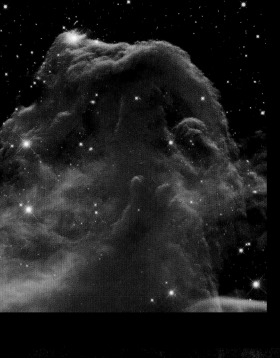

these longer wavelengths, where the absorption of the light is nearly a factor of 10 less than in visible light, nearly all the light from behind the nebula is blocked.

The Horsehead Nebula is part of a much larger cloud known as the Orion molecular cloud complex. This is an area of active star formation, the molecular gas is so thick that the temperature inside the cloud can drop to just a few degrees above absolute zero. This is cold enough to allow the gas to collapse under its own gravity and form new stars. The red glow in visible light is from hydrogen gas being ionized by some of the hot, young stars which have been formed in this region.

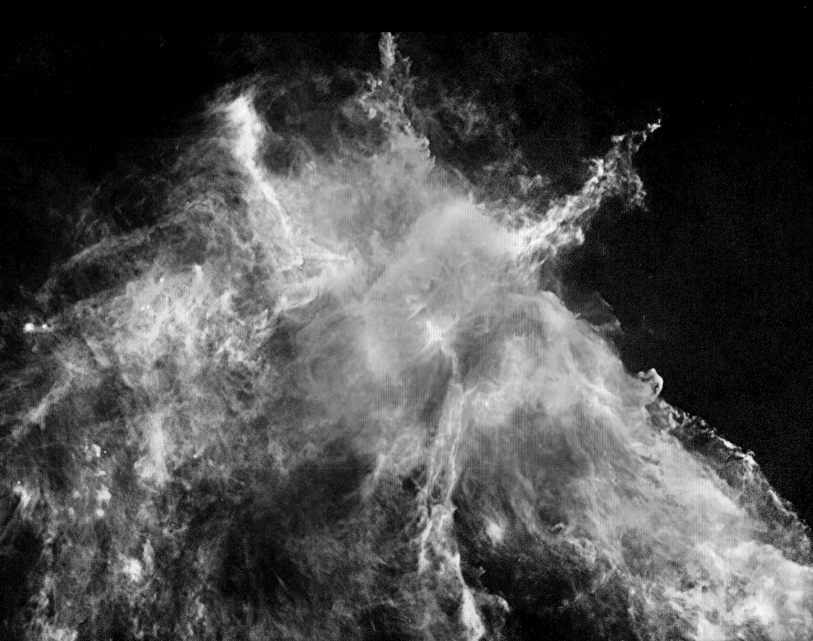

CRAB NEBULA

Very Large Array, Two Micron All-Sky Survey, Spitzer, Hubble, Astro 1 and Chandra
Radio, Microwave, Infrared, Visible Light, Ultraviolet and X-rays

- In 1054 a massive star exploded as a supernova. The supernova became so bright that it was visible in the daytime for the best part of a month. Nearly a thousand years later we see the remnants of this colossal explosion as the Crab Nebula, also known as Messier 1. In fact, the Crab Nebula was the first object to be recognized as a supernova remnant, and at its heart is a rapidly rotating neutron star, the remains of the star which exploded all those years ago.

The gas thrown off by the exploding star has been expanding ever since, it is currently measured to be moving outwards at about 1,500 kilometres/second. At the nebula's centre is the neutron star, rotating just over 30 times every second. This neutron star is also a pulsar; it emits pulses of radiation from gamma rays to radio waves. These pulses of radiation excite the gases in the supernova remnant, leading to the radiation we can see in the images below across the electromagnetic spectrum from radio waves to X-rays.

Estimates of the mass of the gas in the supernova remnant allow us to get an idea of the mass of the progenitor star. Calculations show the mass of gas is about 4.5 times the mass of the Sun. When this is added to the mass of the neutron star, it gives a rather low value for the progenitor star of less than 7 solar masses. We suspect that several solar masses were carried away in fast solar winds before the star exploded.

The Crab has been extensively studied, and the six images below show it across a wide range of wavelengths from low-energy radio waves, which trace electrons spiralling at near the speed of light along magnetic field lines, to high-energy X-rays, which trace very hot gas at temperatures of millions of degrees.

Right: The Crab Pulsar, a rapidly rotating neutron star of unimaginable density, propels matter and antimatter to near the speed of light in the centre of the Crab Nebula in this image from the Chandra X-ray Observatory and the Hubble Space Telescope.

Radio

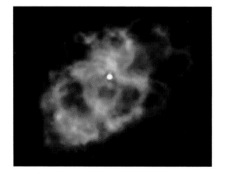

Microwave

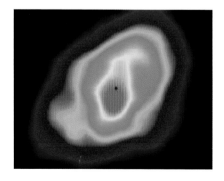

Infrared

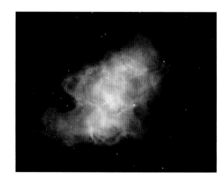

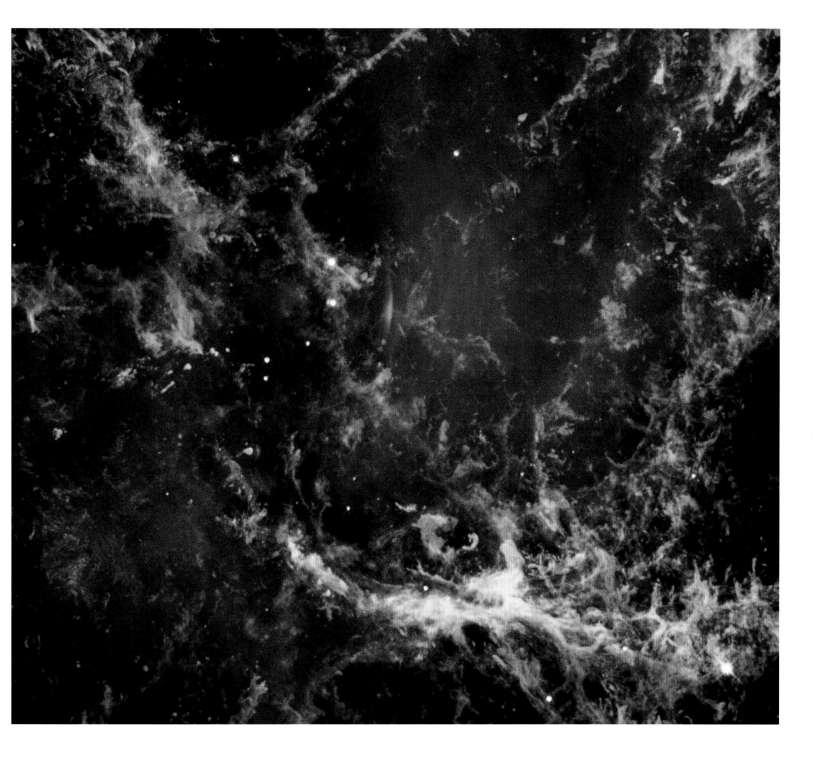

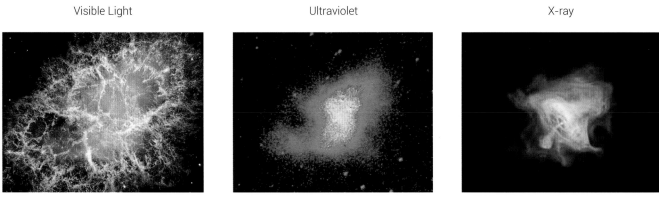

Visible Light   Ultraviolet   X-ray

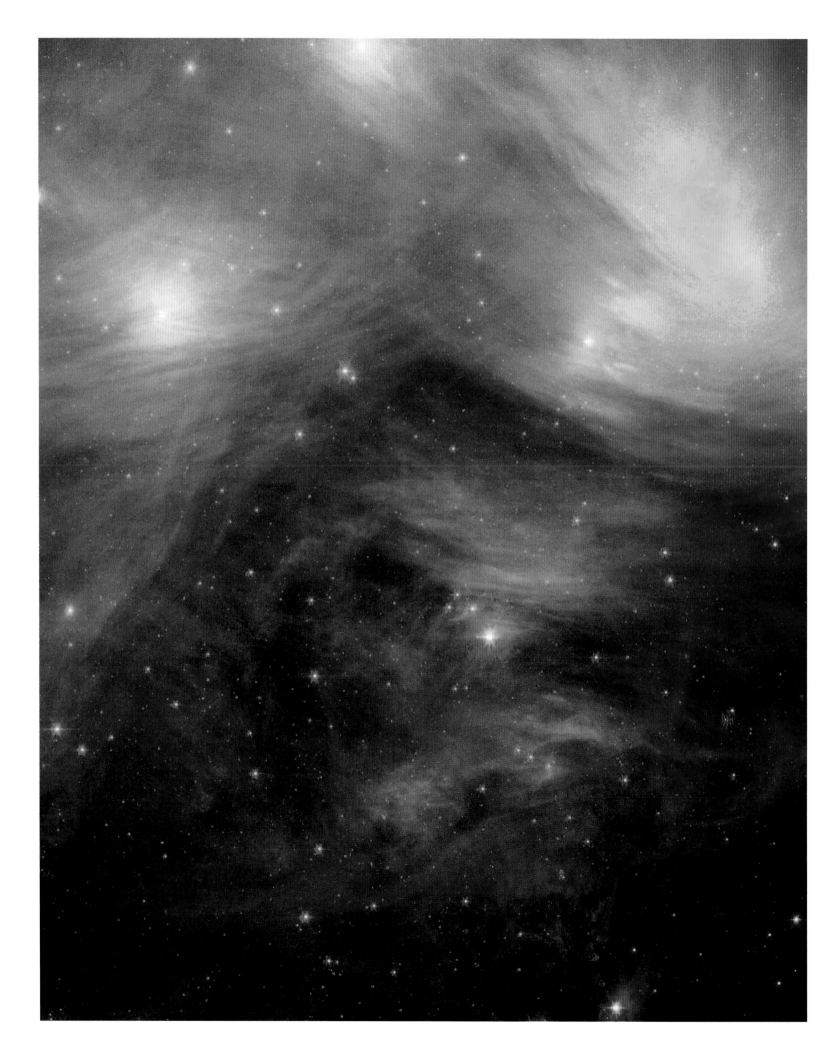

THE PLEIADES (M45)

Spitzer Space Telescope and Wide Field Survey Explorer
Infrared

- The Pleiades, also known as the Seven Sisters or Messier 45, is one of the best known objects in the sky. Lying in the constellation Taurus, it is prominent during the winter months of the Northern Hemisphere. At a distance of about 440 light-years from Earth it is one of the closest star clusters to us, and is calculated to be about 100 million years old.

The infrared image (left) from Spitzer shows a spider's-web-like network of filaments, coloured yellow, green and red, made up of dust, which at visible light wavelengths scatters the starlight, giving the cluster its blue appearance. The densest portion of the cloud appears in yellow and red, and the less dense regions are green hues. Initially it was thought that the dust was left over from the formation of the stars, but we now realize that the cluster is sufficiently old that this dust would have long since dissipated. Instead, it seems that the cluster just happens to be passing through a different cloud of dust, unrelated to its formation.

The image below shows a much larger area infrared view of the Pleiades taken by NASA's Wide-field Infrared Survey Explorer (WISE). The blue and cyan in the image are shorter infrared wavelengths at 3.4 and 4.6 microns and show emission mainly from stars. The green and red represent light at 12 and 22 microns and show emission from warm dust. This WISE infrared image shows the true extent of the dust cloud through which the star cluster is passing; it is much larger than implied in familiar visible light images.

Below: The Pleiades as imaged in the infrared by NASA's WISE telescope.

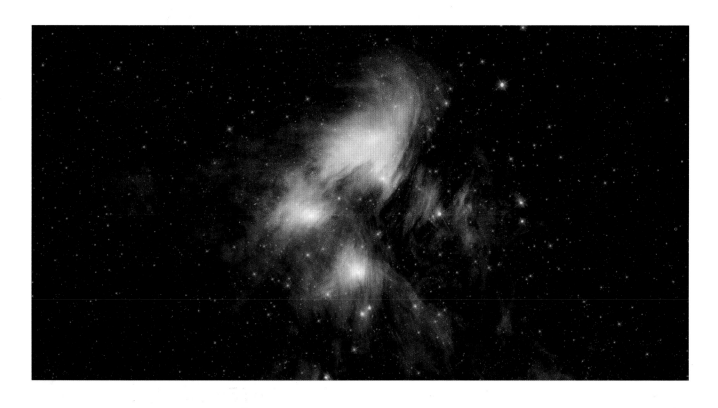

THE STELLAR NURSERY IC 2944

Very Large Telescope
Visible Light

• To celebrate 15 years of operation, in 2013 ESO's Very Large Telescope took this stunning view of the stellar nursery known as IC 2944. Also known as the Running Chicken Nebula, IC 2944 is to be found near the star Lambda Centauri in the southern constellation Centaurus. It is an example of an open cluster, similar to the Pleiades shown on page 85 – but unlike the Pleiades where, in visible light, we see reflection of starlight from dust, in IC 2944 we see emission from the gases which surround the newly formed stars.

The characteristic red glow comes from ionized hydrogen gas; the energetic ultraviolet photons from the newly formed stars are able to strip the hydrogen atoms of their single electron. When these liberated electrons recombine with protons they cascade down through various electron energy levels in the hydrogen atoms. The red light seen here is when an electron jumps down from the third energy level to the second energy level, and is known as H-alpha emission.

Also visible in this image are a number of dark patches. These are thick clouds of dust which go by the name of Bok globules, which are often the sites of star formation as their cold, shielded environment provides an ideal place for clouds of gas to collapse under their own gravity. However, it is believed that these particular globules, known as Thackeray globules after South African astronomer David Thackeray who discovered them in 1950, will not have time to form stars because they are being eaten away by the same ultraviolet photons which are ionizing the hydrogen gas.

Below: General view of the Paranal Observatory Platform with six domes (from left to right): ANTU, KUEYEN, MELIPAL, AT1, VLT Survey Telescope and YEPUN.

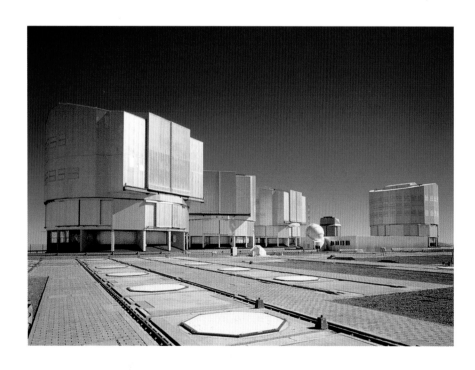

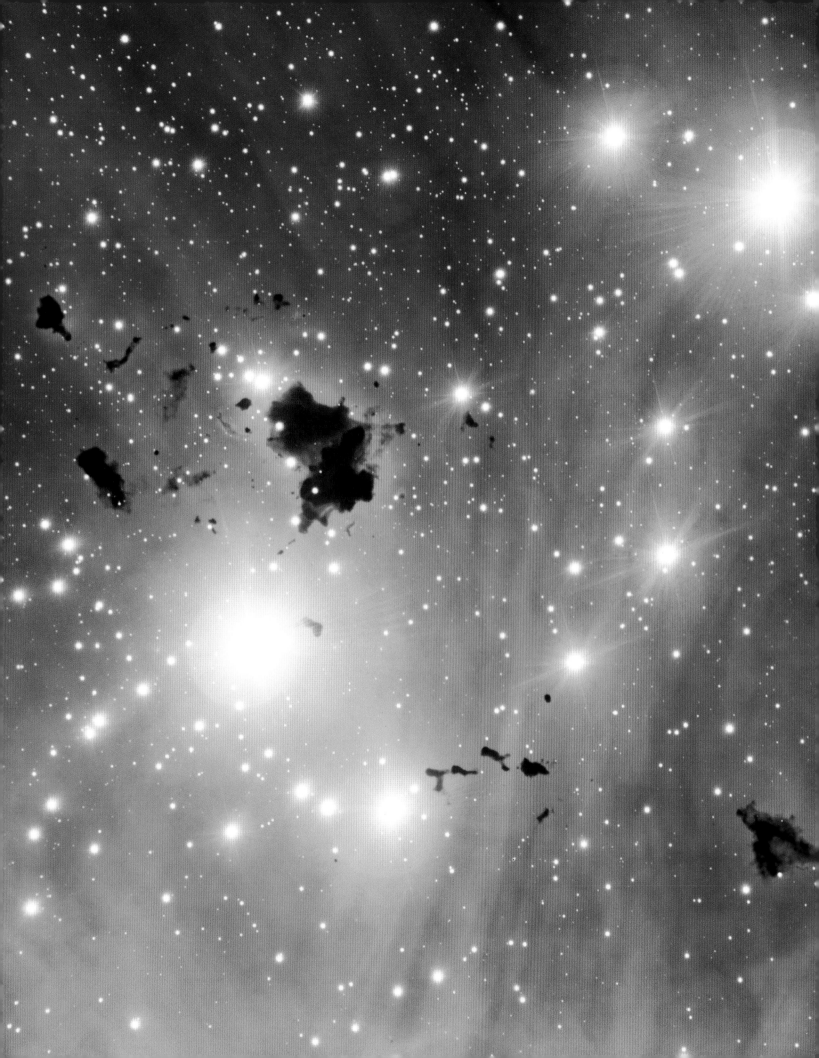

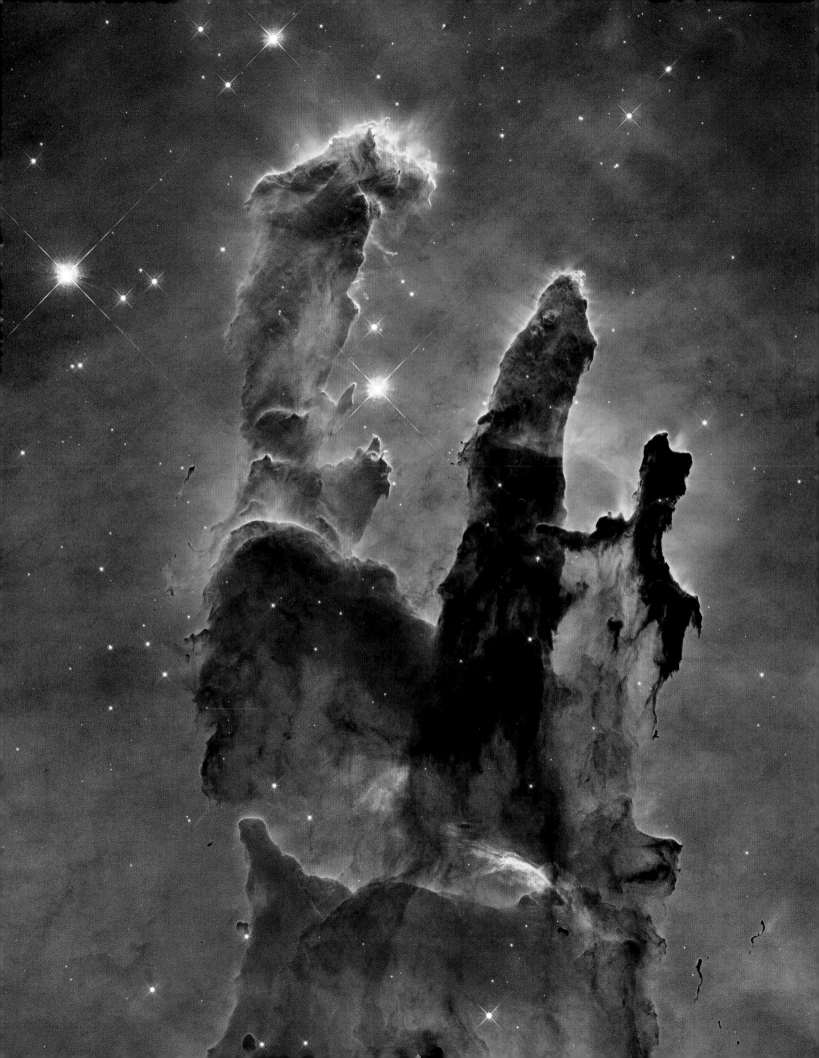

PILLARS OF CREATION

Hubble Space Telescope
Visible Light and Infrared

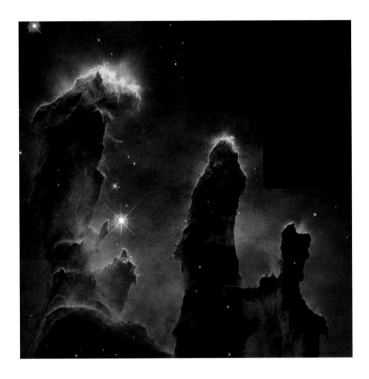

- In 1995 the Hubble Space Telescope took one of its most iconic images. The image is of part of the Eagle Nebula (Messier 16), and became known as the Pillars of Creation. It shows filaments of dense molecular clouds where new stars are being formed. Nearly 20 years later, in 2014, Hubble revisited the nebula and took a new image (opposite) which shows a number of interesting differences.

  The original image was taken using Hubble's Wide Field and Planetary Camera 2 (WFPC2), a camera which was put on Hubble during the 1993 mission to correct for the error in Hubble's primary mirror. What we see in the WFPC2 image is emission from various gases in the cloud; the green is from hydrogen, the red from singly ionized sulphur and the blue from doubly ionized oxygen.

  When Hubble revisited the Pillars of Creation in 2014 WFPC2 had been replaced by the Wide Field Camera 3 (WFC3), which was installed in 2009. WFC3 obtained a larger area image, but if you look closely you will also notice a few other important differences. The new image has a higher resolution than the original, it has a better dynamic range, and it is also a combination of visible and infrared light. These are all technical differences, but if you look very closely at the filaments you can see that they have actually altered their structure in the intervening twenty years; the pillars are slowly being eaten away by high energy radiation from some of the stars.

Above: The original Hubble Space Telescope visible light image of the Pillars of Creation, taken in 1995.

SERPENS STAR-FORMING CLOUD

Spitzer Space Telescope and Two Micron All-Sky Survey
Infrared

- New stars in the process of forming are shown in this infrared image. The image is of an object known as the Serpens Cloud Core, a star-forming region which lies about 750 light-years from Earth in the constellation Serpens. One of the things which makes this star-forming cloud interesting is that it only contains stars of low to moderate mass; very different from the Orion Nebula (see page 71) which is dominated by emission from very luminous massive stars.

  To image this star-forming region observations from two separate telescopes have been combined. The longer wavelength infrared data are from the Spitzer Space Telescope and shorter wavelength infrared data are from the 2 Micron All-Sky Survey (2MASS), a ground-based astronomical survey which imaged the whole sky using telescopes in Arizona and Chile. The three separate infrared wavelengths have been assigned red, green and blue colours in this combined image.

  This entire star-forming cloud is hidden from view at visible wavelengths. But, in the infrared we are able to peer through much of the dust and see inside the cloud. The infrared image shows the presence of young stars in orange and yellow, and shown in blue we also see in the centre a cloud of gas. Off to the left of the centre of the image is a dark patch, this is a part of the cloud which is enshrouded in so much dust that not even the infrared light used here can penetrate it. Most of the stars shown in this image are not in the cloud; they lie either in front of or behind the Serpens Nebula.

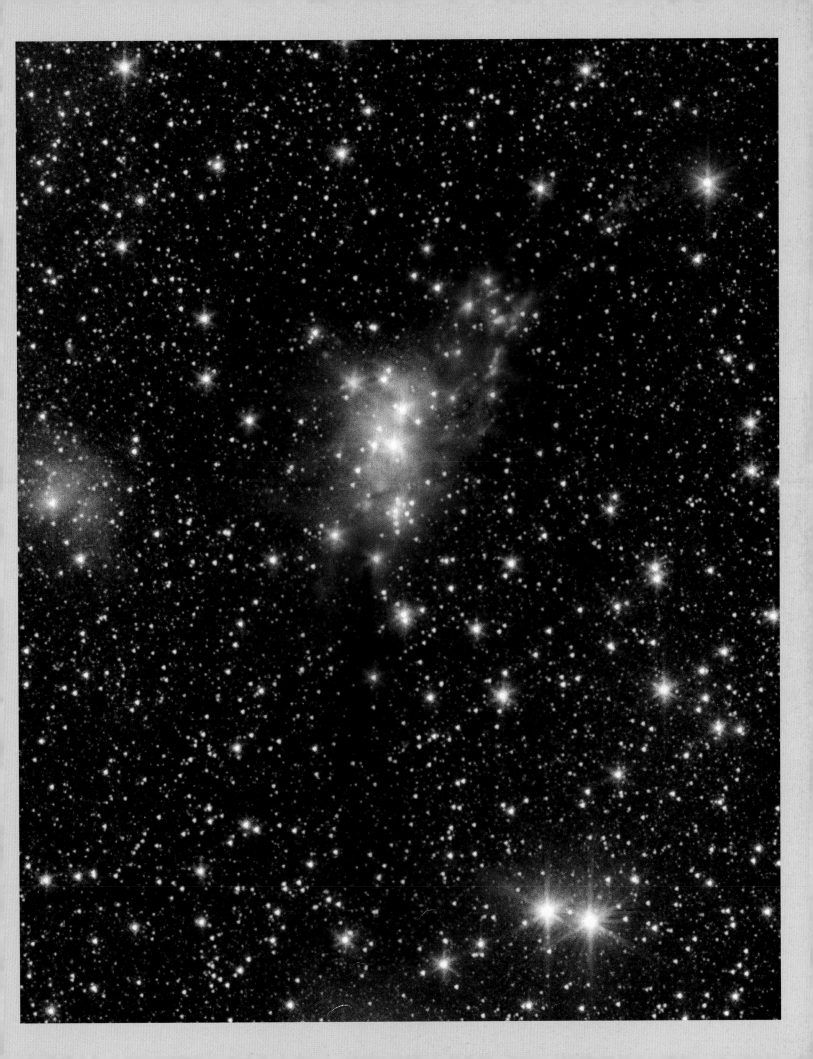

SUPERMASSIVE BLACK HOLE

Hubble and Spitzer Space Telescopes and Chandra X-ray Observatory
Infrared, Visible Light and X-rays

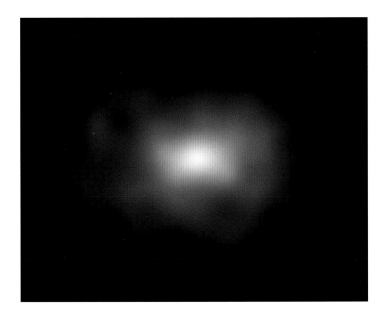

- The motion of stars near the centre of our Milky Way shows the presence of a supermassive black hole. Two groups, one based in Germany and the other in the United States of America, have monitored the orbits of individual stars near the radio source Sagittarius A*, a bright radio source believed to lie at the centre of our Galaxy. The two groups' observations have spanned a period of nearly two decades.

  If one can determine the size of orbit of a body and how long the body takes to orbit, it is possible from the laws of gravity to calculate the mass of the object lying at the centre of the body's orbit. When this was done for a number of stars in motion about Sagittarius A*, it was found by the two groups that the stars were orbiting a mass of more than four million times the mass of the Sun. This mass was found to be enclosed within a sphere with a diameter of only 44 million kilometres (for comparison, the diameter of the Earth's orbit is about 300 million kilometres). The most likely explanation for an object of such a high mass within a relatively small volume is a supermassive black hole.

  Studies of other galaxies have led to the suggestion that nearly all spiral and elliptical galaxies harbour a supermassive black hole at their centres. For a handful of galaxies that we know contain a supermassive black hole, a tight correlation has been found between the mass of the black hole and the dispersion in the velocities of the stars in the bulges of those galaxies. It is argued that the formation of the black hole and of the galaxy itself are linked in some fundamental way.

Above: A close up shows Sagittarius A* (Sgr A*) only in X-rays, covering a region half a light-year wide. The X-ray emissions are from hot gas which is being sucked in to the black hole.

Left: The supermassive black hole Sgr A* at the centre of our Milky Way Galaxy is located in the bright, white region at the right of this composite image from NASA's great observatories Chandra, Hubble and Spitzer.

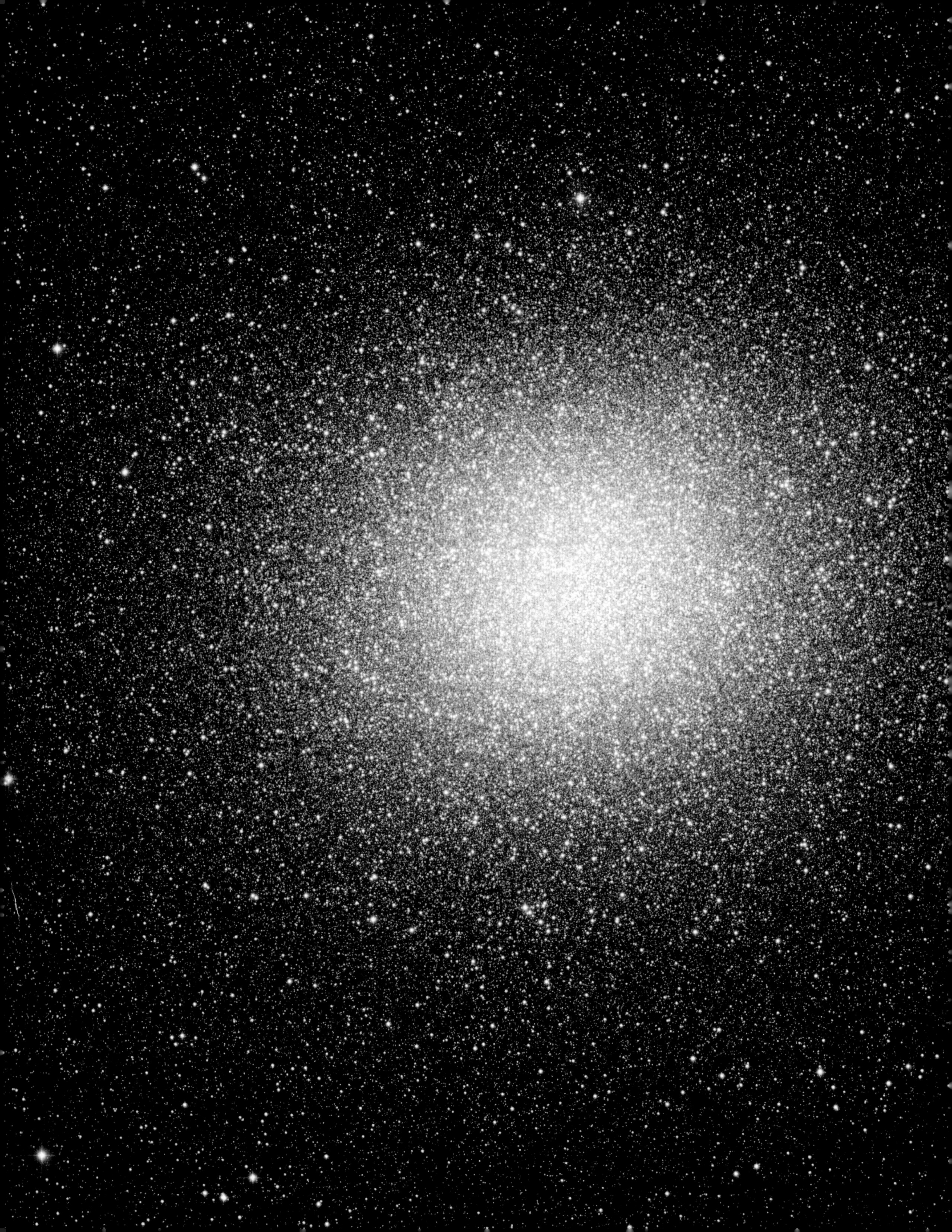

OMEGA CENTAURI

Very Large Telescope Survey Telescope
Visible Light

- The largest globular cluster in the Milky Way Galaxy is Omega Centauri. For many centuries it was believed to be a star; the Greco-Egyptian astronomer Ptolemy described it as "a star on the horse's back". In 1603 German cartographer Johann Bayer designated it as Omega Centauri, one of the fainter stars in the constellation Centaurus. However, observing through a telescope from the island of Saint Helena in the Southern Atlantic in 1677, English astronomer Edmond Halley realized that it was non-stellar, and in 1715 he included it in his list of six "luminous spots or patches".

  It was first recognized as a globular cluster by Scottish astronomer James Dunlop in 1826. It is one of a few globular clusters which are visible to the naked eye; from a dark location it appears about the same size as the full Moon, half a degree across in the sky. Located nearly 16,000 light-years from us, it has a diameter of roughly 150 light-years and contains approximately 10 million stars with a total mass of some four million times the mass of the Sun. It has been calculated to be about 12 billion years old.

  The stars in the core of Omega Centauri are estimated to lie only 0.1 light-years from each other (our Sun's nearest neighbour is just over four light-years away). Its size and distinctive appearance have led many to argue that Omega Centauri is the core remnant of a dwarf galaxy that has been stripped of its material during numerous passes through the disk of our Milky Way.

PART 3

EXPLORING THE
LOCAL GROUP

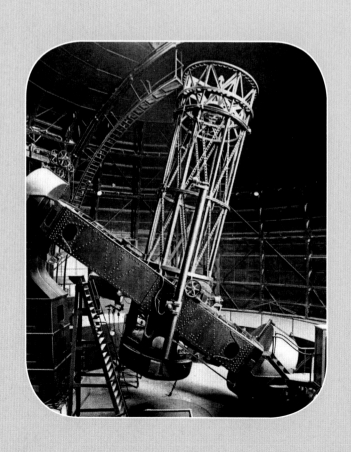

The Local Group is our corner of the Universe, our neighbourhood. A few of its members are visible to the naked eye; our Milky Way, the Andromeda Galaxy, and the Large and Small Magellanic Clouds. The vast majority of its members, however, can only be seen through telescopes and several have only recently been discovered. The Local Group's most distant member, the Pinwheel Galaxy (Messier 33), lies just under three million light-years from us. Andromeda, the most distant object visible to the naked eye, lies just over two million light-years away.

Until the 1920s it was hotly debated whether our Milky Way Galaxy was the entire Universe. With increasingly powerful telescopes, other spiral nebulae similar in appearance to the Andromeda Nebula were discovered, and the debate arose as to whether these were vast stellar systems beyond our Milky Way, or clouds of gas in the process of forming stars within our own Galaxy. The matter was finally resolved in 1923, when Edwin Hubble used the 100-inch Mount Wilson Telescope to measure the distance to the Andromeda Nebula, calculating a distance which was far too big for it to be part of our Galaxy.

In the decades since then, our understanding of our neighbourhood has continued to improve. The Local Group is dominated by two large spiral galaxies, our Milky Way and the Andromeda Galaxy. Together, these two galaxies account for more than half the mass. There are also a number of smaller isolated galaxies, such as the Pinwheel Galaxy and dwarf galaxies such as Sextans A. Most of the remaining members of the Local Group that we have so far discovered are in orbit about our Milky Way, these include the Large and Small Magellanic Clouds.

The Local Group is now known to be an outlying part of the Virgo Cluster, a large collection of galaxies which we discuss in the next section. In this section we present images of some of the Local Group's members, including dwarf galaxies in orbit about our Milky Way which have only recently been discovered. Many of the images shown here were obtained by the Hubble Space Telescope, but other visible light data have been provided by large ground-based telescopes such as the European Southern Observatory's Very Large Telescope. Data at other wavelengths come mainly from space-based telescopes, such as infrared images taken by the Spitzer Space Telescope and X-ray images taken by the Chandra X-ray Observatory. As with our Milky Way, observations at other wavelengths enable us to properly understand the vast array of environments which we find in the varied members of our Local Group.

Opposite: Inside the dome of the 100-inch Mount Wilson Reflecting Telescope on a 1,740-metre peak in the San Gabriel Mountains near Pasadena, northeast of Los Angeles.

HIGH VELOCITY GAS CLOUDS

Hubble Space Telescope
Visible Light

- One of the more remarkable findings about our Milky Way Galaxy in the last few decades has been the discovery of high velocity clouds of molecular hydrogen gas (right). These high velocity clouds (HVCs) are measured to have velocities in excess of 70–90 kilometres/second, they can have masses in excess of millions of times the mass of the Sun, and they cover large portions of the sky. HVCs are found in the halo of our Milky Way, and have also been observed in other nearby galaxies such as the Andromeda Galaxy.

  HVCs play an important part in our understanding of the evolution of our Galaxy. They account for a large fraction of the amount of normal matter in our galactic halo, and it is thought that they add material to the disk of our Galaxy as they fall in. This new material helps maintain the star formation rate in the Galaxy which would otherwise have used up the gas present in the disk.

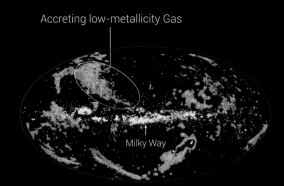

Accreting low-metallicity Gas

Milky Way

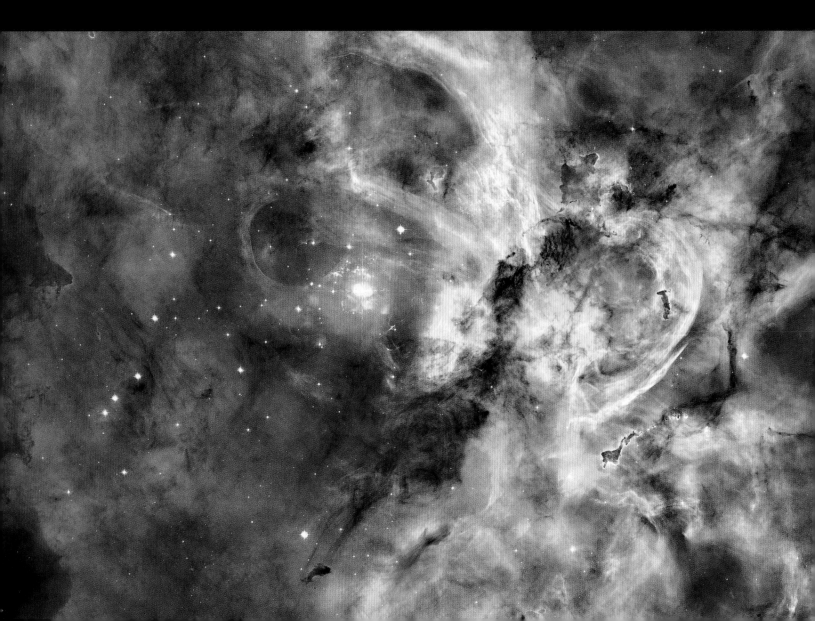

The origin of the HVCs is still a subject of vigorous debate. Some of the HVCs are undoubtedly due to the interaction of our Galaxy with its satellite galaxies. For example, gravitational interaction between the Milky Way and the Large and Small Magellanic Clouds has produced the well-known Magellanic Stream. But, the origin of other HVCs is less well understood, and so there are probably other origins for some of the HVCs which have nothing to do with interactions with satellite galaxies. One possibility is that some of the HVCs are gas which was ejected from the Milky Way during an earlier epoch and is now falling back towards the disk, the so-called Galactic Fountain.

In the very detailed Hubble image of the Carina Nebula below, the dark figures are molecular clouds, knots of molecular gas and dust which are so thick that they have become opaque.

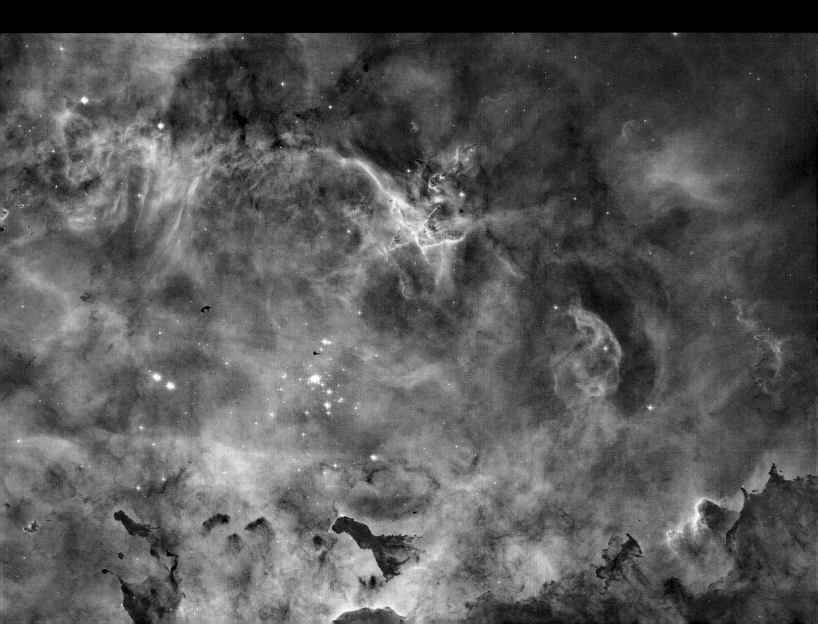

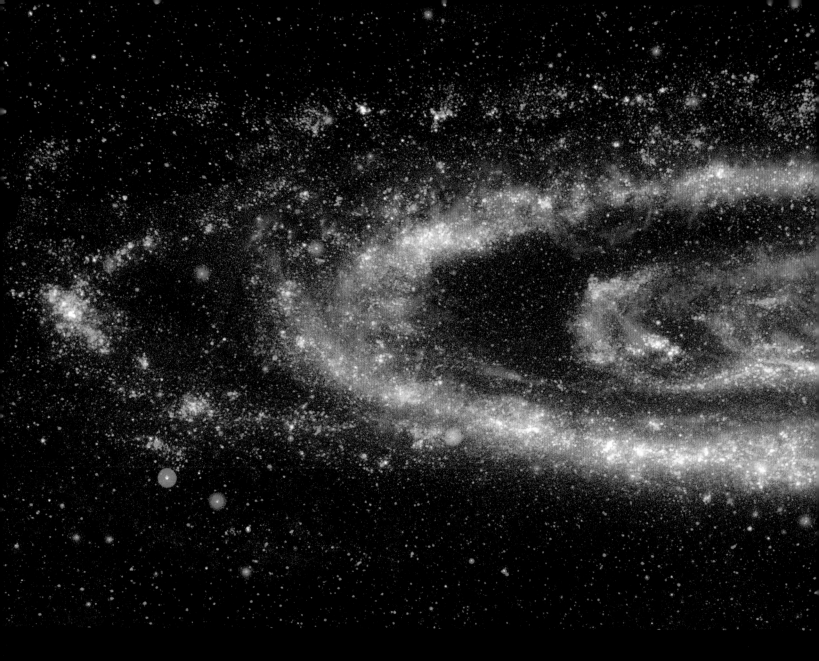

ANDROMEDA GALAXY M31

Spitzer Space Telescope and Galaxy Evolution Explorer
Infrared and Ultraviolet Light

- In 1923 Edwin Hubble was obtaining photographs of the Great Andromeda Nebula (Messier 31) using the 100-inch telescope at Mount Wilson Observatory in Southern California. One night he noticed three new stars, and marked them with an "N" (for "nova") on his photographic plate. After studying some images that he had taken previously he realized that one of them was a variable star, so the "N" was crossed out and replaced with a "Var!". The exclamation mark was because Hubble realized that he could use the period-luminosity relation for Cepheid variables to determine the distance to this nebula. When he did, our understanding of the Universe changed; he found it was far too distant to be part of our Milky Way Galaxy. The Andromeda Galaxy, as we now call it, is our closest large galaxy. It is also the

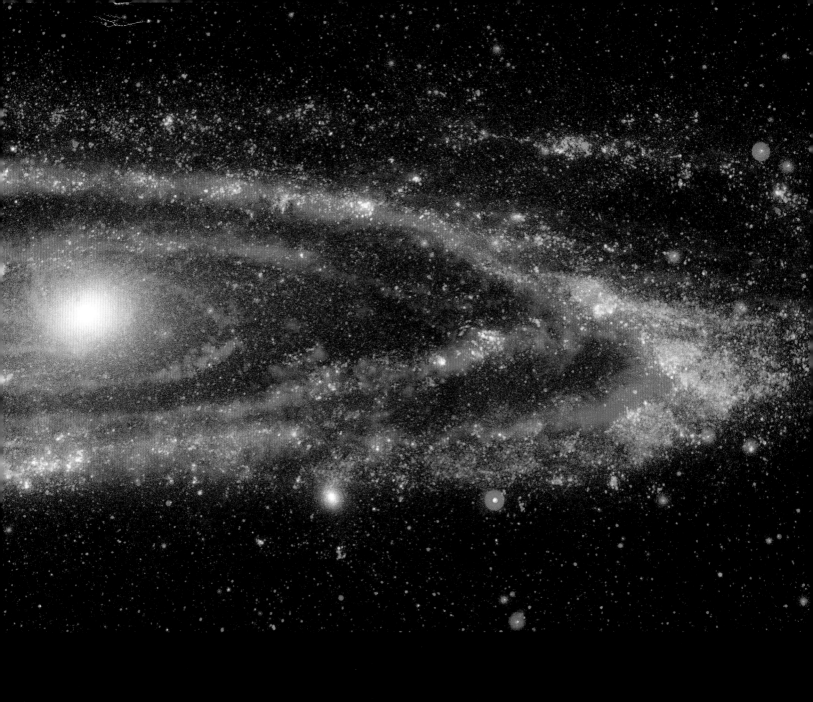

most distant object visible with the naked eye, lying at a distance of a little over two million light-years away. Being our nearest large galaxy, it is also the most studied large galaxy.

In this composite image NASA's ultraviolet satellite GALEX (Galaxy Evolution Explorer) detected young, hot, high-mass stars, represented in blue, and relatively older stars, shown in green. The central yellow spot represents a population of old stars. Red regions show where the Spitzer Infrared Telescope found cool, dusty, star-forming regions.

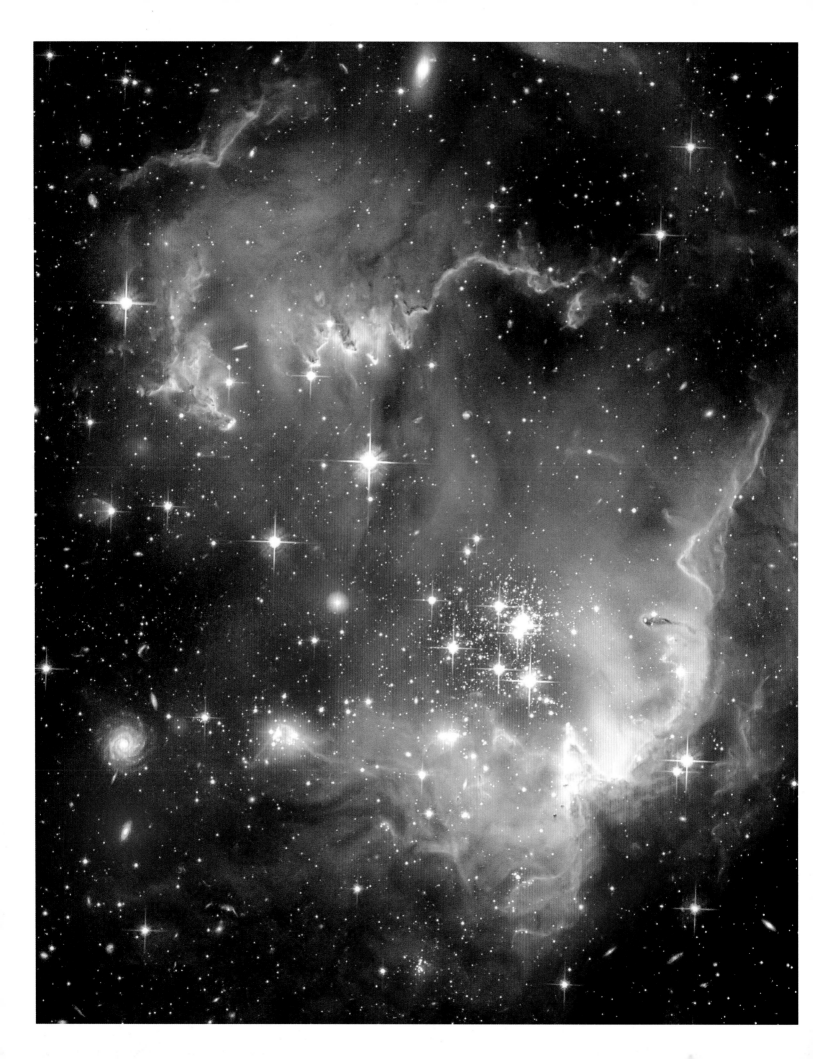

THE SMALL MAGELLANIC CLOUD

Chandra X-ray Observatory, Hubble and Spitzer Space Telescopes
Infrared, Visible Light and X-rays

- Part of the Small Magellanic Cloud (SMC) is shown in this stunning multi-wavelength combined image taken by three of NASA's great observatories. The part of the Small Magellanic Cloud which has been imaged is known as NGC 602, and this cluster is of particular interest to astronomers, as it lies in the wing of the SMC which leads to the Magellanic Bridge. Because it lies at the periphery of the SMC, NGC 602 is easier to study than many other parts of this dwarf irregular galaxy.

In the image presented here, NGC 602 has been observed by the Chandra X-ray Observatory (shown in purple), in visible light by the Hubble Space Telescope (shown in red, green and blue), and in infrared light using the Spitzer Space Telescope (also shown in red). In addition to NGC 602, a number of background galaxies can be seen towards the edges of the image.

Because the SMC and the Large Magellanic Cloud (LMC) are comparatively close to us, they allow us to study phenomena which we cannot see in more distant galaxies. For example, this image shows the first time that X-ray emission has been seen in young stars, with masses similar to the mass of our Sun, outside of our own Galaxy. The infrared emission recorded by Spitzer shows the presence of large quantities of dust which is associated with gas, providing the raw material for new episodes of star formation. In fact, some of the most actively star-forming regions anywhere in the Local Group are found in the SMC and LMC.

The SMC played a key role in our understanding of the scale of the Universe. In 1908 Henrietta Leavitt discovered thousands of variable stars in photographs taken at the Harvard College's observatory in Chile. Many of these variable stars belonged to a particular class of variables known as Cepheid variables. By comparing photographs taken over many nights, Leavitt noticed that the brighter variables in the SMC took longer to vary their brightness compared to the fainter ones. She reasoned that, as the stars all lay in the SMC, she could assume that they were all at essentially the same distance. By 1912 she had established a relationship for these Cepheid variable stars which we now know as the period-luminosity relationship.

Following pages: The Small Magellanic Cloud seen in visible light by the 2.2-metre Max Planck Gesellschaft Telescope at the La Silla Observatory in Chile.

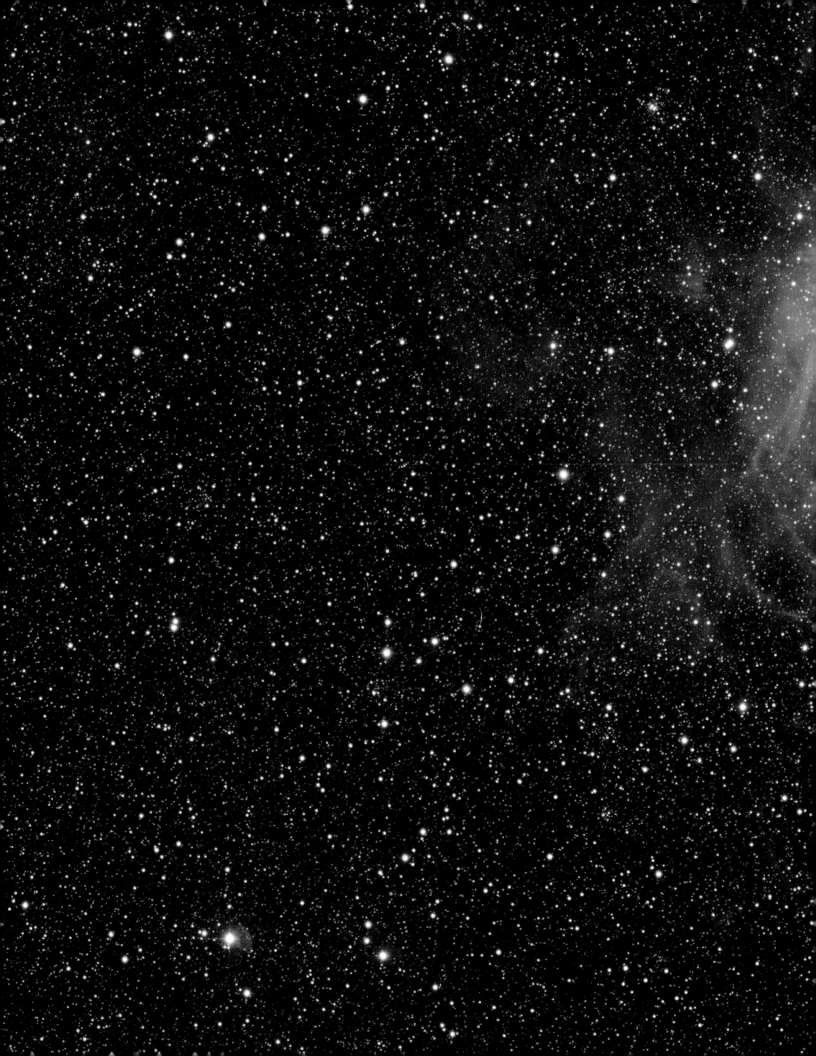

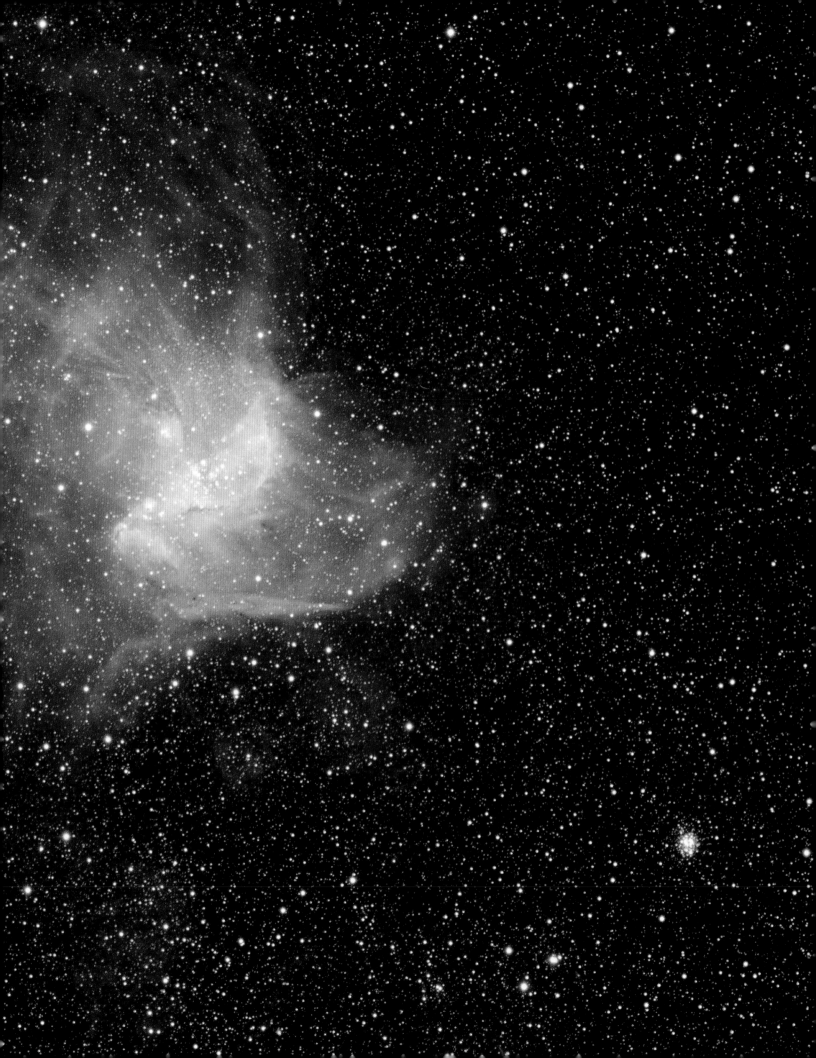

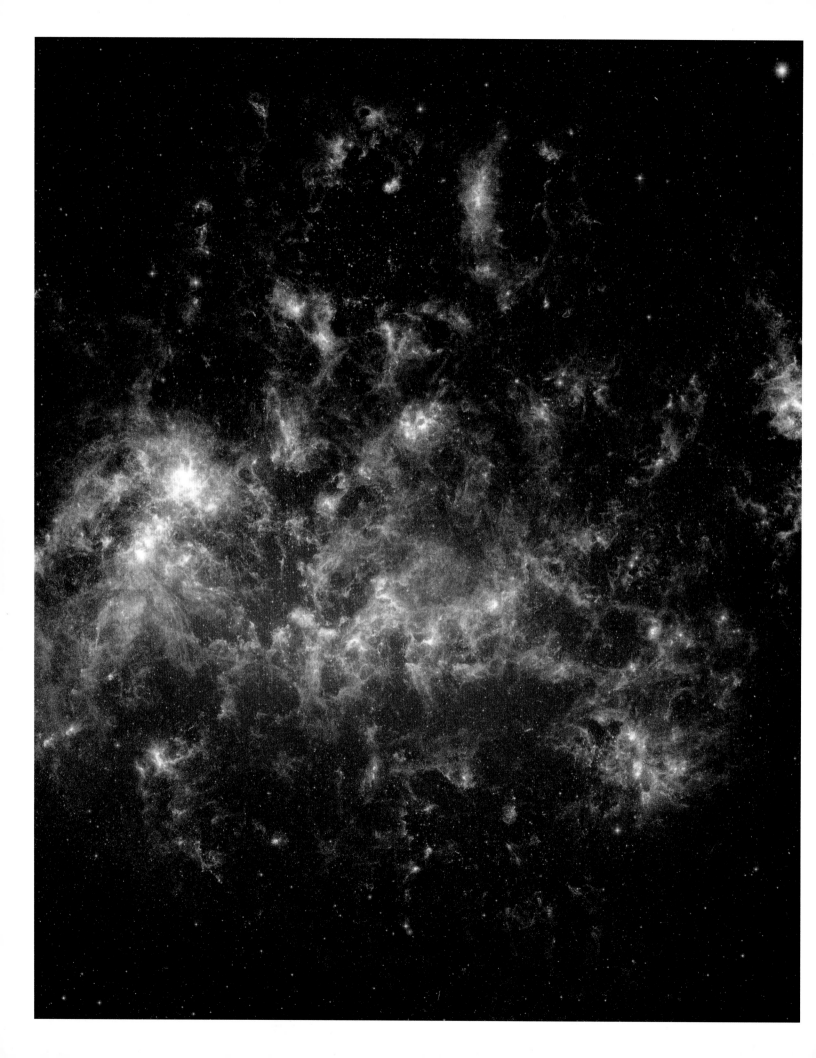

THE LARGE MAGELLANIC CLOUD

Spitzer Space Telescope
Infrared Wavelengths

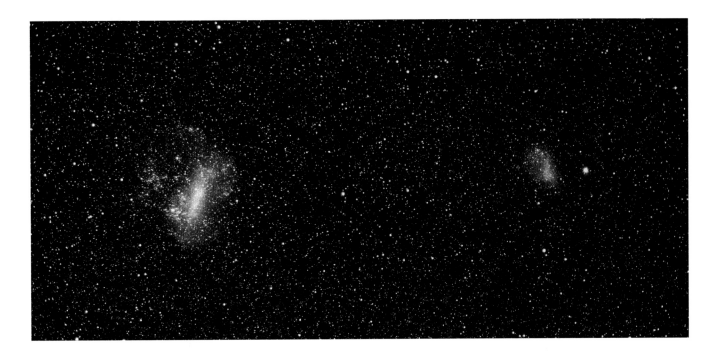

Far to the south in the southern skies lies one of the most remarkable sights in astronomy, the Large Magellanic Cloud (LMC). Along with its smaller companion, the Small Magellanic Cloud (SMC), these two large patches in the sky are visible only south of the Equator. They were first seen by Europeans in the fifteenth century when Dutch and Portuguese sailors noticed them as they sailed around the southern tip of Africa; they were initially known as "Cape Clouds". They were noted again in 1503-4 by Amerigo Vespucci, and during Ferdinand Magellan's circumnavigation in 1519-22 the sailors noted two large "clouds" which were visible from southerly latitudes. They were described in some detail by Antonio Pigafetta who sailed with Magellan, and from this they have become known as the Magellanic Clouds.

The LMC is located some 163,000 light-years away, slightly closer than the SMC which is some 200,000 light-years away. Although not as distant as the Andromeda Galaxy, they are the most distant objects visible to the naked eye in the Southern Hemisphere. The LMC is an example of a disrupted barred spiral galaxy, the bar can be seen in the image above. Its irregular appearance may well be due to tidal interactions with the SMC and our Galaxy. The LMC covers a large area of the sky, about 20 times the diameter of the full Moon. It is rich in gas and dust with many sites of recent star formation, including the famous Tarantula Nebula, the most active star-forming region in the Local Group.

The infrared Spitzer image at left is a mosaic of 300,000 individual tiles. Every stage of the lifecycle of a star is to be found in this single galaxy, which comprises nearly one million objects.

Above: The Large and Small Magellanic Clouds photographed by David Malin at the Anglo-Australian Observatory.

Following pages: This Hubble Space Telescope image shows the Tarantula Nebula in visible, infrared and ultraviolet light. This region is full of star clusters, glowing gas, and thick, dark dust. This image was captured using Hubble's Wide Field Camera 3 (WFC3) and Advanced Camera for Surveys (ACS).

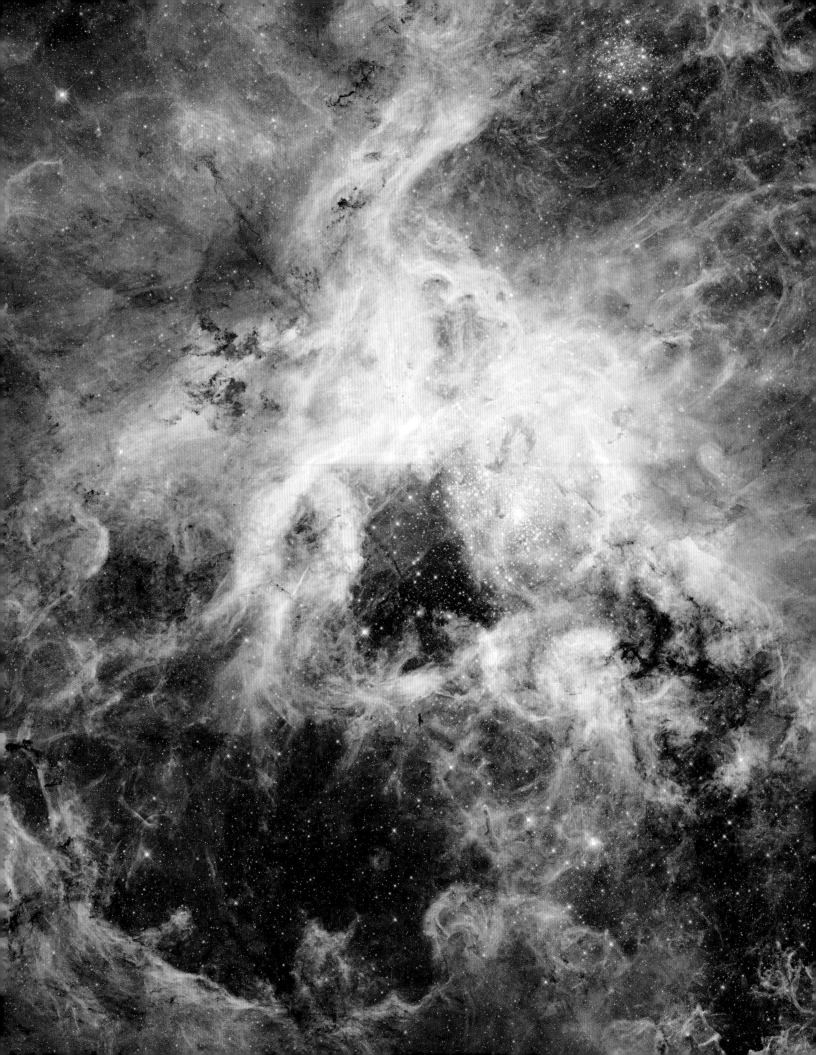

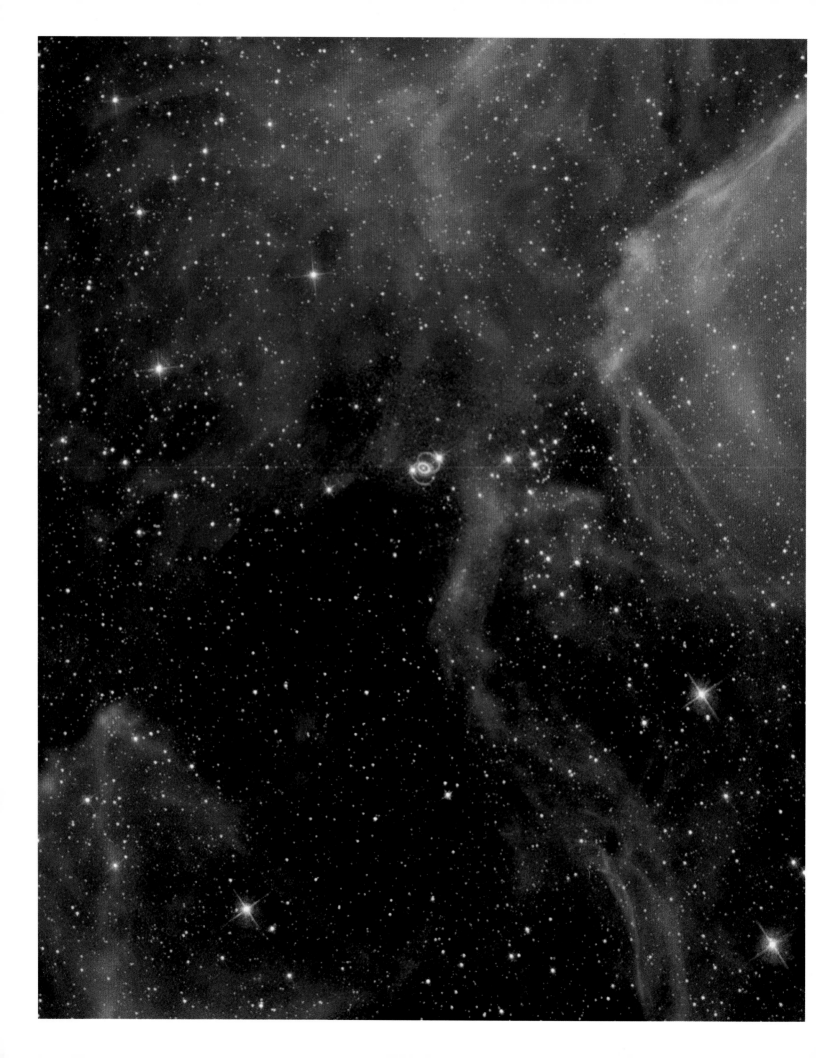

SUPERNOVA 1987A

Hubble Space Telescope
Visible Light

 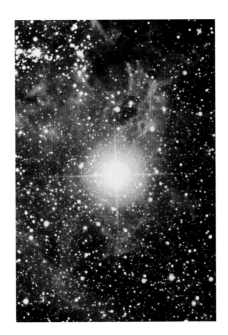

- Just three years after the discovery of Supernova (SN) 1987A, the Hubble Space Telescope (HST) was launched. The closest supernova to have been seen since the invention of the telescope could now be studied with the best telescope available. Over the years, the HST has taken many images of SN1987A, allowing us to study the evolution of this stellar explosion in unprecedented detail. The picture opposite is one of these images, and was taken by the Advanced Camera for Surveys (ACS), which was put on the HST in 2002.

The three bright rings seen in the image at left were discovered by the HST in 1994 and have been studied ever since. The light is from material emitted by the progenitor of SN1987A in stellar winds some 20,000 years before the explosion. This material has been ionized by the ultraviolet flash from the supernova explosion. In about 2001 the ejecta from the supernova explosion, travelling at more than 7,000 kilometres/second, collided with the inner ring. This heated up the gas and caused it to emit X-rays, and the X-ray flux from this inner ring increased by a factor of three between 2001 and 2009. Some of this X-ray emission has been absorbed by the dense ejected gas close to the centre of the supernova remnant, which has led to a similar increase in the visible light brightness of the remnant over the same period.

To date, the HST has failed to see the neutron star expected at the centre of SN1987A. One possibility is that it is enshrouded in dust, another is that it collapsed into a black hole.

Above: David Malin, working at the Anglo-Australian Telescope in New South Wales, Australia, obtained this famous image (above right) of the supernova the night after the explosion was first seen, 24 February 1987. The image (above left) shows the same patch of sky before the supernova.

THE SEAHORSE NEBULA

Hubble Space Telescope
Visible Light

- Looking like a grazing seahorse, the dark object at the bottom right is actually a huge cloud of gas and dust suspended in interstellar space. Located in the Large Magellanic Cloud, a satellite galaxy of our own Milky Way, the Seahorse Nebula (also known as NGC 2074) is about 20 light-years long and is near the star-forming Tarantula Nebula.

This part of the LMC is undergoing a huge burst of star formation, possibly triggered by a nearby supernova explosion. Such stellar explosions send shock waves into the surrounding gas clouds, and can cause cool molecular hydrogen clouds to collide and form a new generation of stars. The Universe is an excellent recycler; the material thrown off in a supernova explosion is not just used in a future generation of stars, but the explosion helps the formation of these new stars happen.

The visible light image shown here is approximately 100 light-years across; we see dark clouds of gas and dust rising above the main molecular cloud, illuminated by emission from ionized gases in the thinner parts of the gas cloud. The high-energy ultraviolet radiation responsible for ionizing the gases is also eating away at the surface of the dense molecular cloud. This image was taken in 2008 by the Wide Field Planetary Camera 2 on the Hubble Space Telescope. Red shows emission from ionized sulphur atoms, green from ionized hydrogen and blue from ionized oxygen.

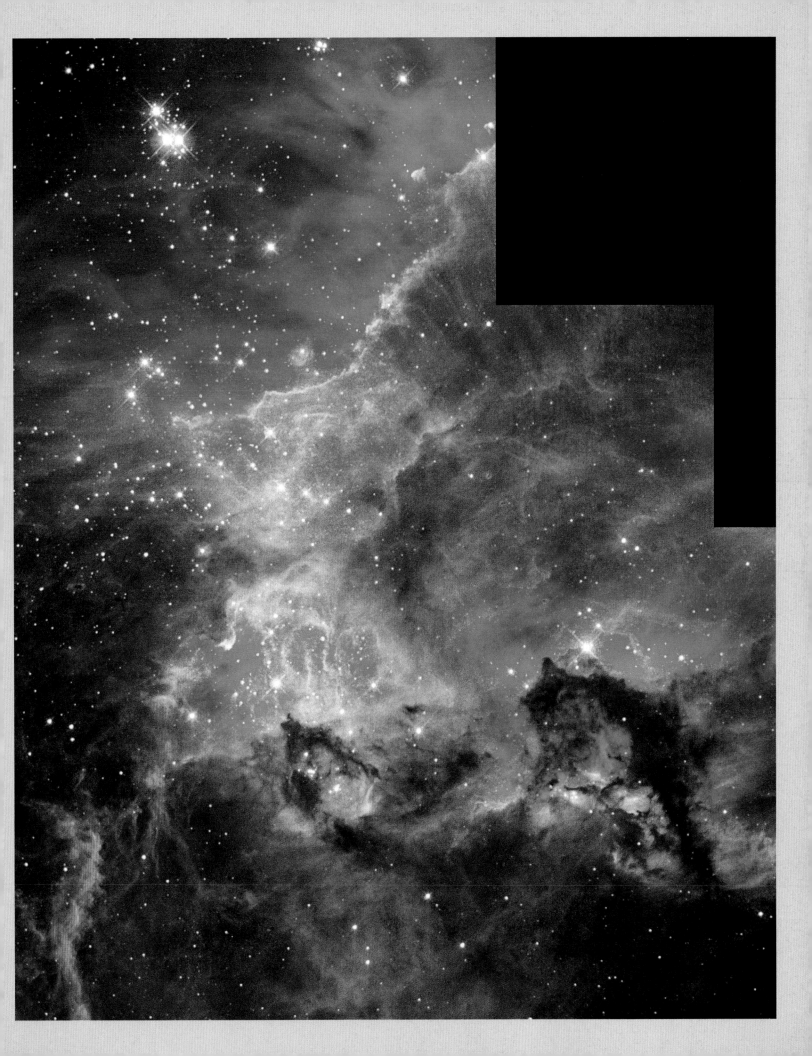

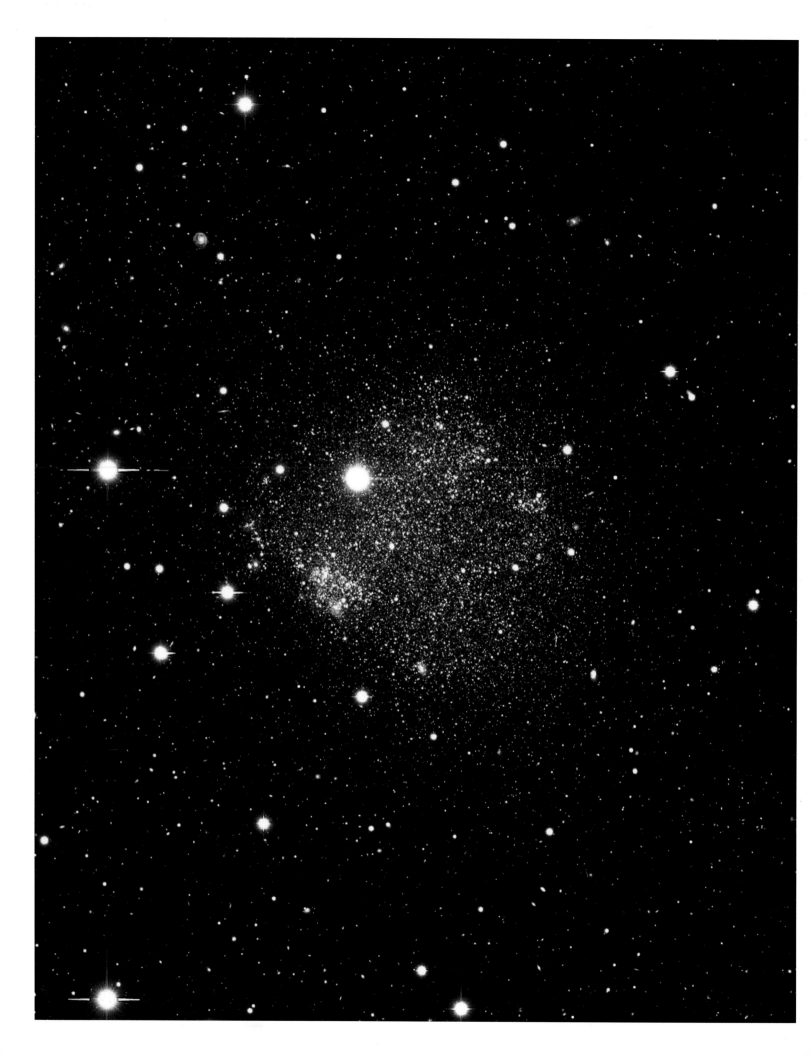

SEXTANS A DWARF GALAXY

Subaru Telescope
Visible Light and Near-Infrared

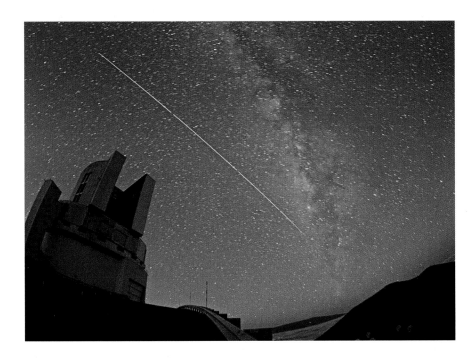

- Sextans A is a tiny dwarf galaxy that is part of our Local Group. It is an example of a dwarf irregular galaxy, showing neither an elliptical nor a spiral structure. Dwarf irregular galaxies often have a high content of gas, the raw materials for creating new stars. Typical of its class, Sextans A exhibits some vigorous massive star formation. These areas of massive star formation appear as blue patches in this image, which was taken by Japan's 8-metre Subaru Telescope located at an altitude of over 4,000 metres on the summit of Mauna Kea in Hawaii.

  The image is a combination of long exposures taken through three separate filters. The first filter is in the green part of the visible spectrum, the second one in the red, and the third just beyond the red end of visible light in the near-infrared. These three separate images have been combined to produce this real colour image.

  Apart from the few regions of massive star formation, Sextans A is an example of a low surface brightness galaxy. This means that the amount of light it emits per unit area is less than is typical for a galaxy. This makes it extremely hard to see; only long exposures through large telescopes are able to reveal its faint structure. Sextans A is only some 5,000 light-years across, a twentieth the diameter of our Milky Way Galaxy. It is also one of the most distant members of the Local Group, lying a little over four million light-years away, about twice the distance of the Andromeda Galaxy.

Above: The International Space Station passing over the Subaru Telescope.

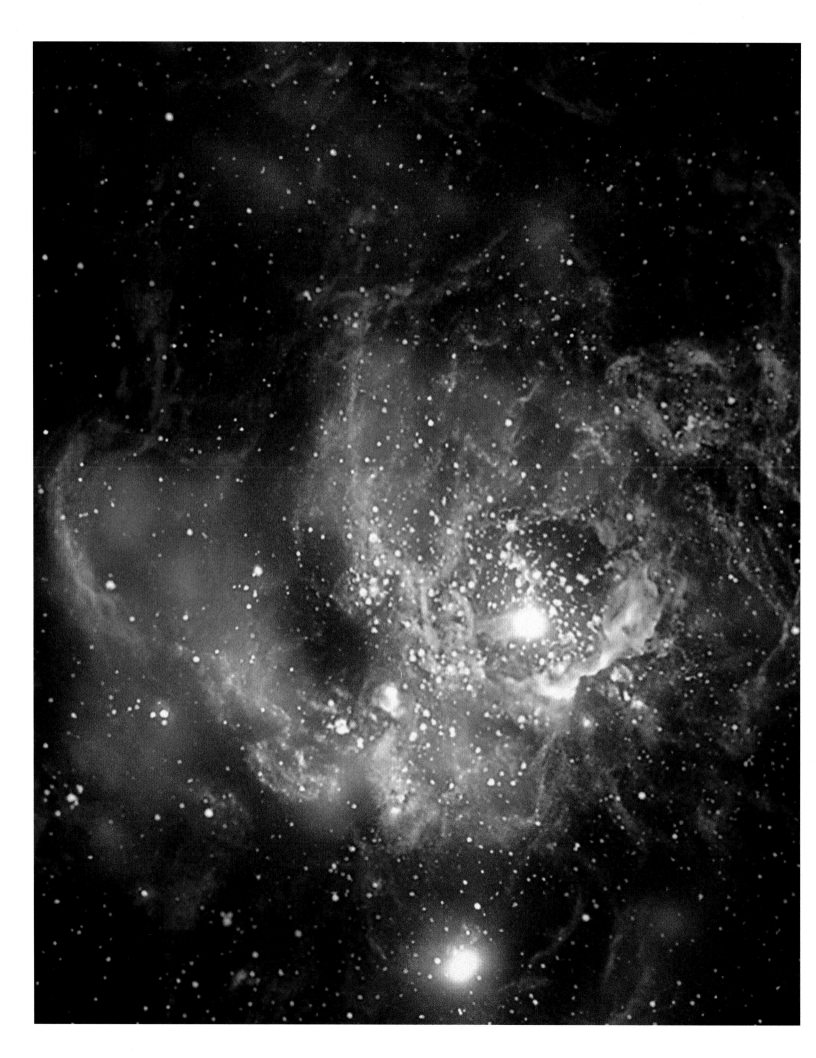

STAR FORMATION IN MESSIER 33

Hubble Space Telescope and Chandra X-ray Observatory
Visible Light and X-rays

- M33 Is the third largest galaxy in our Local Group, lying a little further away than Messier 31 (the Andromeda Galaxy) at a distance of about 2.5 million light-years. It is also known as the Triangulum Galaxy because it lies in the northern constellation Triangulum. Messier 33's spiral structure harbours one of the largest star-forming regions known, NGC 604. This star-forming region's diameter spans almost 1,500 light-years; and we have been able to count more than 200 hot, young stars in the nebula with masses ranging from 15 to over 60 times the mass of the Sun.

Such regions of massive star formation are often known as HII regions, because of the presence of ionized hydrogen which astronomers designate with an "H" for hydrogen and Roman numeral II to show that it is singly ionized. The image below shows the location of NGC 604 in one of the spiral arms of Messier 33.

The main image shows a composite Hubble Space Telescope and Chandra X-ray Observatory study of NGC 604. The visible light HST data are shown in red and green, the Chandra X-ray data are shown in blue. Giant bubbles in the cool dust and warm gas are created by the stellar winds of the hot, young stars being formed in this stellar nursery. But, these same stellar winds are believed to cause some of the gas to collide with the surrounding gas and dust, heating this gas to millions of Kelvin and leading to the X-ray emission which Chandra is able to observe.

Below: The NGC 604 star-forming region in Messier 33 is the bright patch of light seen at the end of one of the spiral arms in the top left quarter of the picture. The individual stars with diffraction spikes are foreground stars in our Milky Way. (Image from the Palomar Sky Survey.)

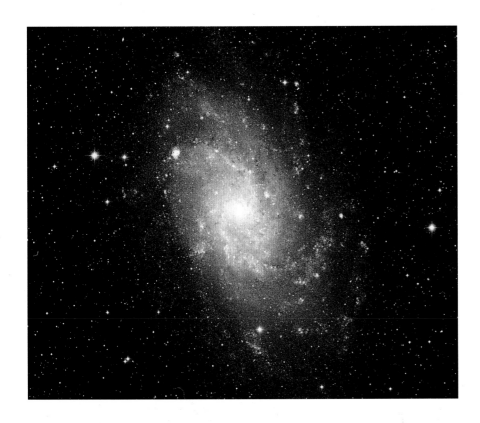

PART 4

BEYOND THE LOCAL GROUP

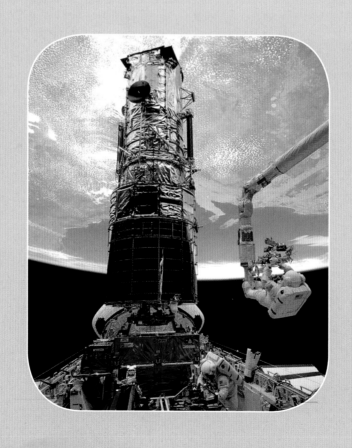

The Universe beyond the Local Group shows a richness and variety far greater than anything that we see in our neighbourhood. Whereas our Local Group is dominated by two large spiral galaxies, beyond the Local Group we see a much greater variety of galaxies and environments. Galaxies are seen in many shapes and sizes; giant elliptical galaxies, majestic spiral galaxies, galaxies violently colliding and galaxies with huge bursts of star formation are all part of the plethora of galaxies we can see. We even see strange galaxies whose appearance has been distorted by the light passing close to foreground galaxies, so-called gravitational lenses which were predicted by Einstein but have only recently been seen.

Clusters of galaxies, first observed in the 1930s, are now known to be an essential component of the structure of the Universe. Our nearest cluster of galaxies is the Virgo Cluster, a huge collection of at least 1,300 galaxies whose centre lies about 55 million light-years from us. Some of the better known objects shown here are members of the Virgo Cluster, particularly the galaxies that are part of the Messier catalogue. In turn, the Virgo Cluster is part of an even larger structure known as the Virgo Supercluster, with our Local Group being an outlying part of this supercluster.

With distances from us of tens if not hundreds of millions of light-years, the galaxies and clusters of galaxies shown in this section require the largest telescopes to study them properly. Not only do we require the largest telescopes, but also the most sensitive instruments. Photographic plates, used for over 100 years to capture images of the night-time sky, have since the 1980s been replaced by electronic cameras that are about one hundred times more sensitive. These have allowed us to capture images which would just not have been possible in the past.

In addition to images obtained by the Hubble Space Telescope and the European Southern Observatory's Very Large Telescope, the four 8-metre telescopes high in the Atacama desert in northern Chile, many of the images shown here have been obtained by large telescopes like Japan's 8-metre Subaru Telescope on the summit of Mauna Kea. Other wavelength data have been obtained by various space-based telescopes, including the Herschel Space Observatory and the Compton Gamma Ray Observatory.

These huge telescopes capture light which has often taken tens of millions, and sometimes billions, of years to reach us; some of the light having left their objects before our planet even existed.

Opposite: The first servicing mission to the Hubble Space Telescope in 1993 saw astronauts install a set of specialized lenses to correct the flawed main mirror in the telescope.

MESSIER 81

Hubble and Spitzer Space Telescopes and Galaxy Evolution Explorer
Infrared, Visible and Ultraviolet Light

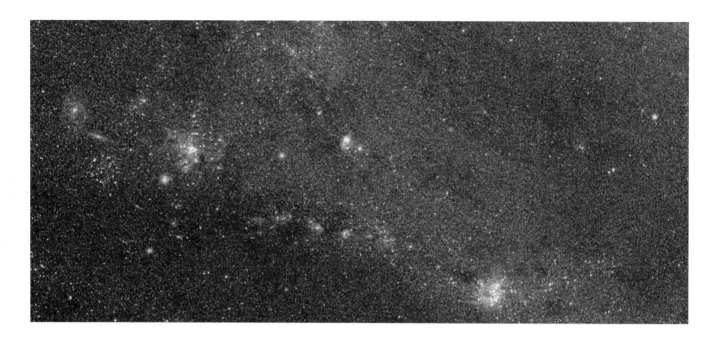

The spiral galaxy Messier 81 is one of the best known extra-galactic objects in the sky. It is an example of what is known as a "grand design" spiral galaxy, meaning that its spiral arms extend all the way towards its centre. Because it is located in Ursa Major, it is visible to Northern Hemisphere astronomers throughout the year. This, combined with its large size and relatively high brightness, make it one of the most popular targets for amateur astronomers. At about 12 million light-years, it is one of the nearest large spiral galaxies to us, making it one of the most studied at various wavelengths.

Messier 81 was discovered by Johann Elert Bode in 1774, and is sometimes known as Bode's Galaxy. Its other common designation is NGC 3031. In addition to the clearly visible spiral arms, M81 is an example of a galaxy with an active nucleus, and has been found to harbour a supermassive black hole at its centre with a calculated mass of 70 million times the mass of the Sun.

The image shown opposite is a combination image showing visible light data from the Hubble Space Telescope (shown in yellow-white), infrared light from the Spitzer Space Telescope (shown in red), and ultraviolet light captured by NASA's Galaxy Evolution Explorer (shown in blue). Ultraviolet light is emitted by the hottest, youngest stars, which are found in the galaxy's spiral arms. Older stars lie in the bulge at the centre of M81, and dust emission is also found along the spiral arms, showing the sites of current and future star formation.

Above: Close up of one of the spiral arms from the Hubble image overleaf showing individual star-forming regions.

Following pages: This Hubble Space Telescope view of M81 is so sharp that it can resolve individual stars, along with open star clusters, globular clusters, and even glowing regions of fluorescent gas.

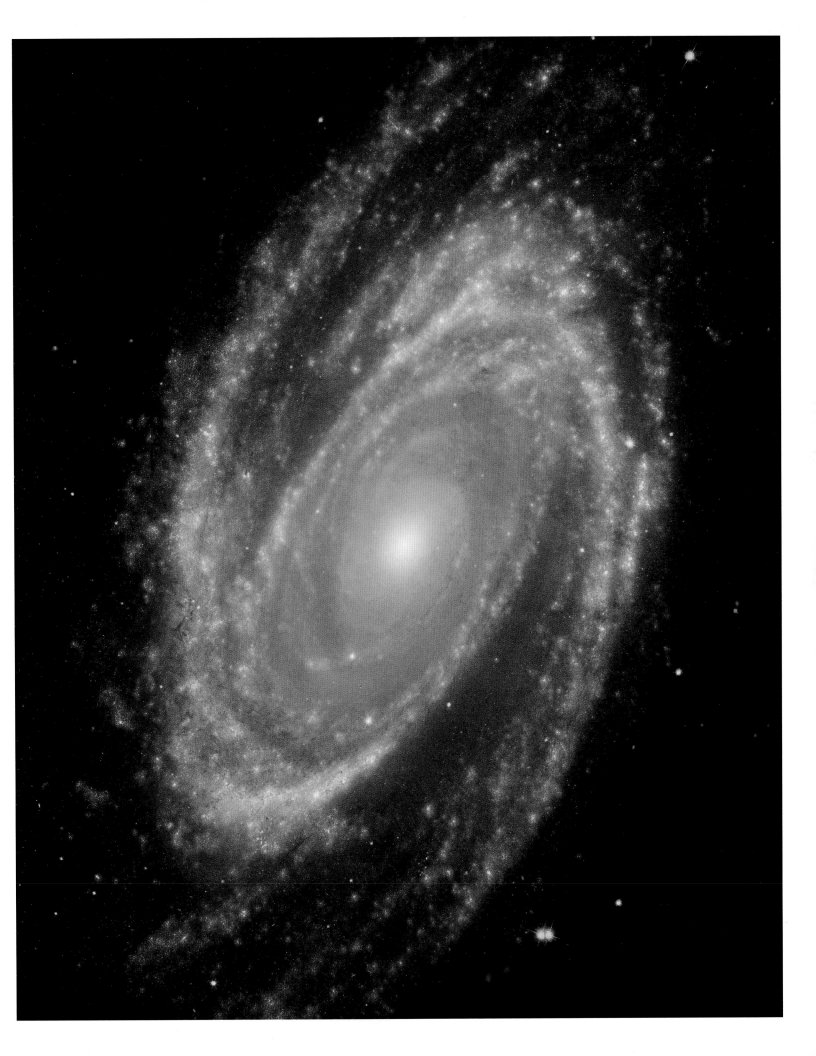

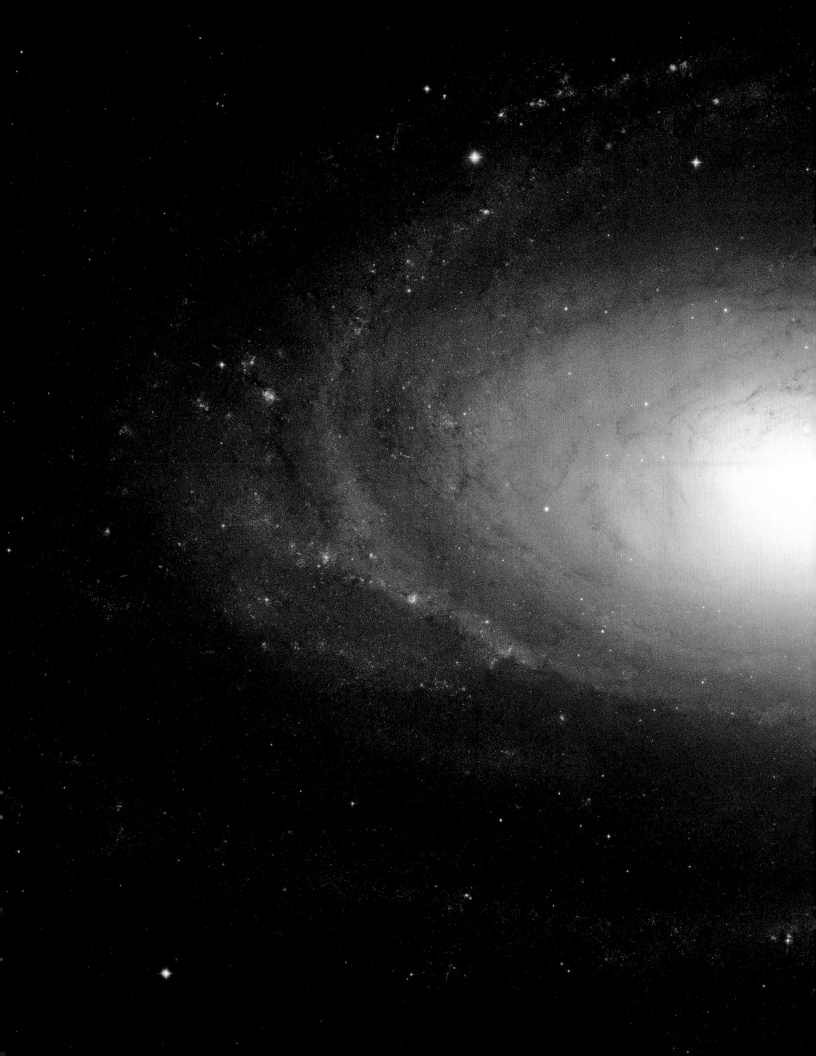

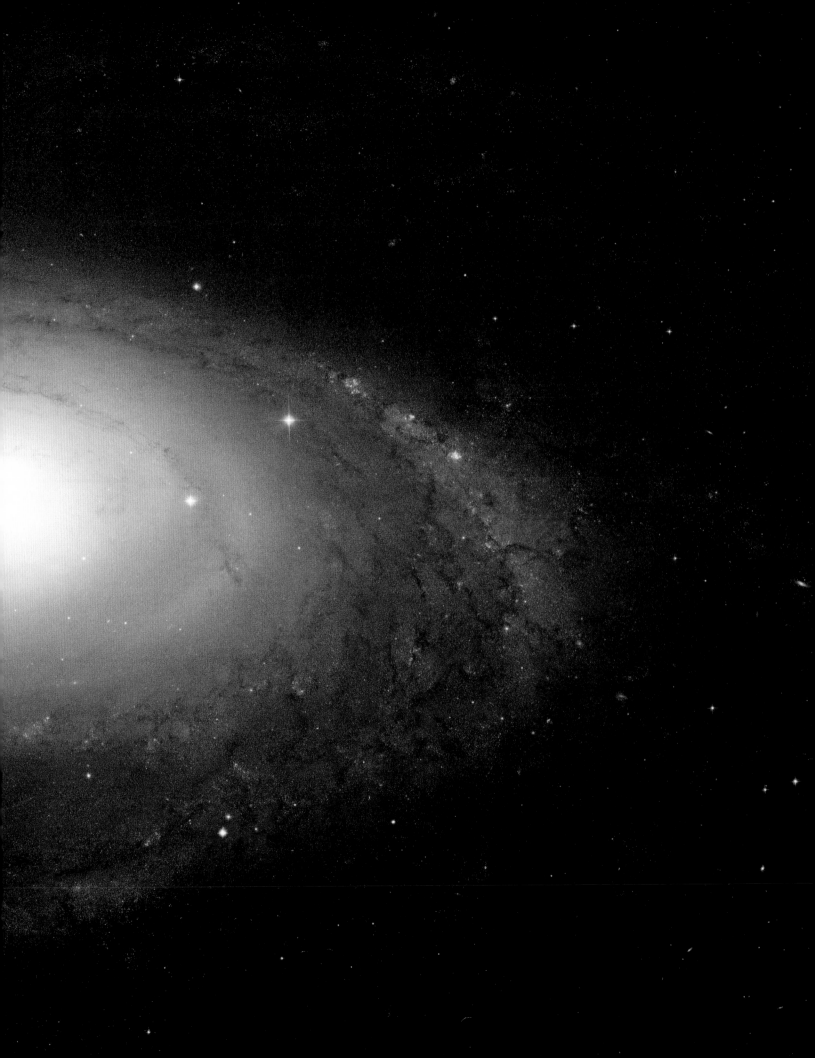

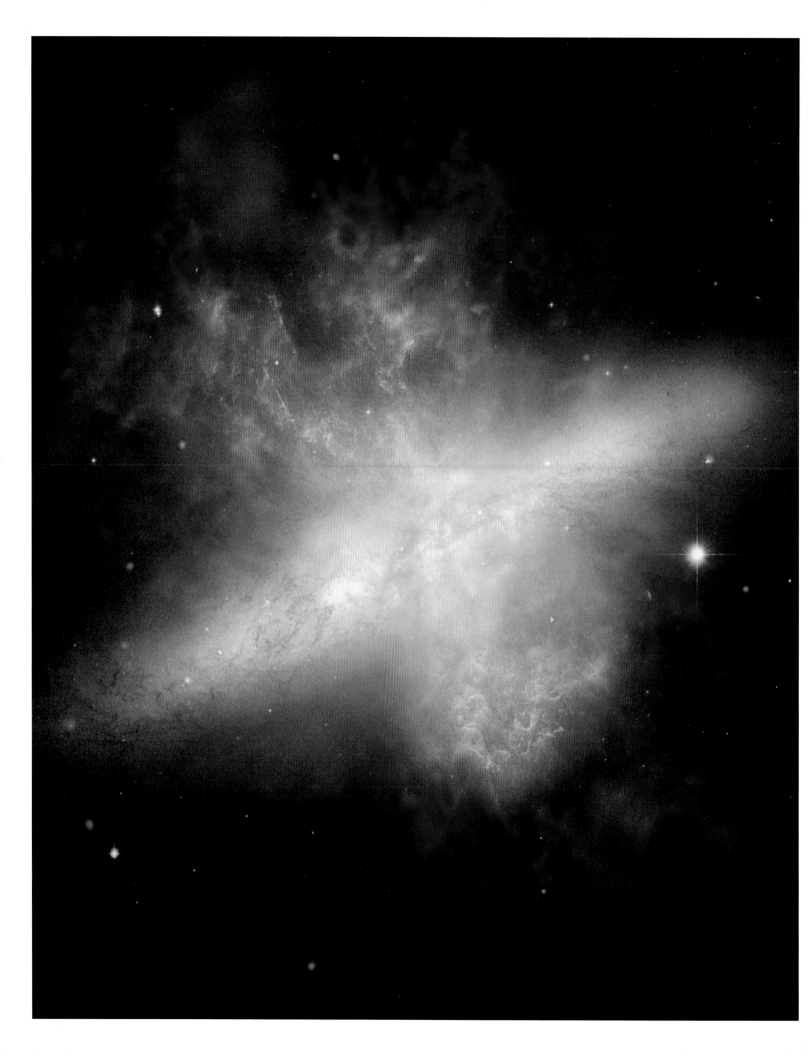

THE CIGAR GALAXY

Spitzer and Hubble Space Telescopes and Chandra X-ray Observatory
Infrared, Visible Light and X-ray

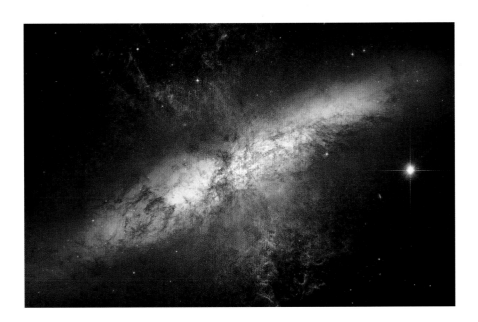

- Messier 82 is one of the brightest and easiest to find galaxies in the sky. It is about five times more luminous than the Milky Way, and at its centre it is undergoing a huge burst of star formation. This burst of star formation is thought to have been triggered by an interaction with its neighbouring galaxy, Messier 81. At a relatively nearby 12 million light-years, it is the nearest example to us of a starburst galaxy. It lies in the constellation Ursa Major, making it visible all-year round for many people in the Northern Hemisphere.

  Messier 82 is also known as the Cigar Galaxy because of its shape. The picture shown here is a composite image taken using three separate satellites. The visible light image was taken using the Hubble Space Telescope. Overlaid on this is infrared emission taken using the Spitzer Infrared Telescope (shown in red) and X-ray emission taken using the Chandra X-ray space telescope (shown in blue). The infrared emission is due to emission from dust, the X-ray emission is due mainly to emission from fast-moving electrons.

  As can be seen from the image, the infrared and X-ray emissions are much more extended than the visible parts of the galaxy. This tells us that the massive burst of star formation at the centre is propelling material and high-energy electrons at right angles to the galaxy. At visible wavelengths, the intense star formation is largely hidden from view due to all the dust in the inner parts of the galaxy. It is only by studying Messier 82 at multiple wavelengths that we can learn its true nature.

Above: Hubble Space Telescope image showing bright starburst clumps in the central parts of the galaxy.

THE SOMBRERO GALAXY

Spitzer and Hubble Space Telescopes
Infrared and Visible Light

- One of the last objects in the famous Messier catalogue of astronomical objects compiled by the nineteenth-century French astronomer Charles Messier is Messier 104, more commonly known as the Sombrero Galaxy. Its eye-catching appearance is due to the view from Earth of a spiral galaxy with a very prominent bulge nearly edge-on. The dust lane of the disk of the galaxy cuts right across the bulge, but because the galaxy is not quite edge-on we can see the dense concentration of stars at the centre of the bulge.

  This is a composite Hubble visible light image superimposed with an infrared image shown in false colours taken by the Spitzer Space Telescope. If you compare the infrared emission observed by Spitzer to the purely visible light image taken by

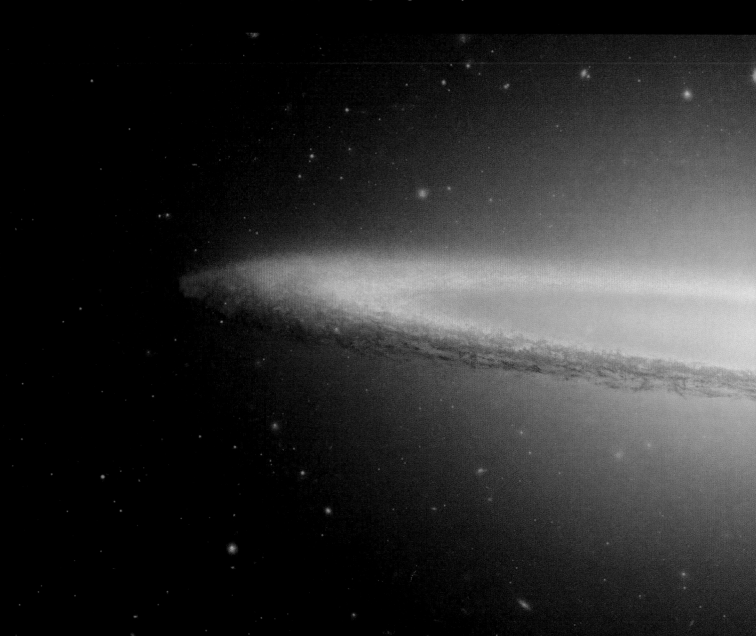

Hubble (overleaf) you can see that the dust which causes the obscuration of the visible light is now seen in emission in the infrared.

The Sombrero Galaxy is in the constellation Virgo, and lies about 30 million light-years from us. It is easily visible through small telescopes, making it one of the most popular targets for amateur astronomers. It is some 50,000 light-years across, about a third the size of our Milky Way, but it is quite different in morphology with a much larger central bulge than our Galaxy has. In addition it has a supermassive black hole at its centre, making it also a very popular target for professional study.

Following pages: The Hubble Space Telescope's original visible light image of the Sombrero Galaxy. Images were taken in three filters (red, green, and blue) to yield a natural-colour image.

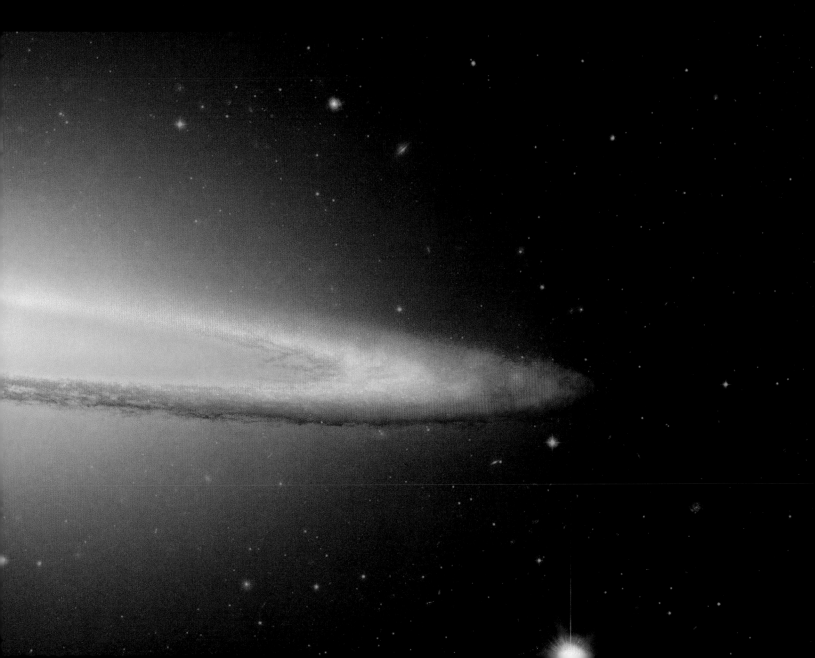

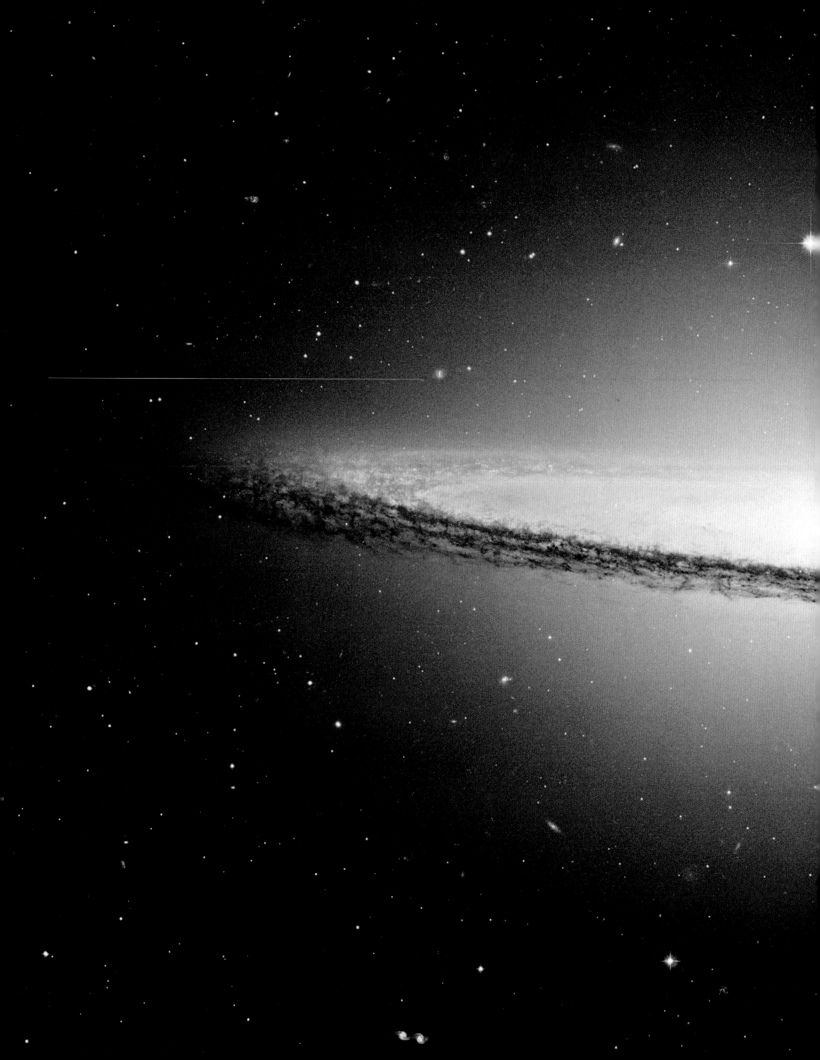

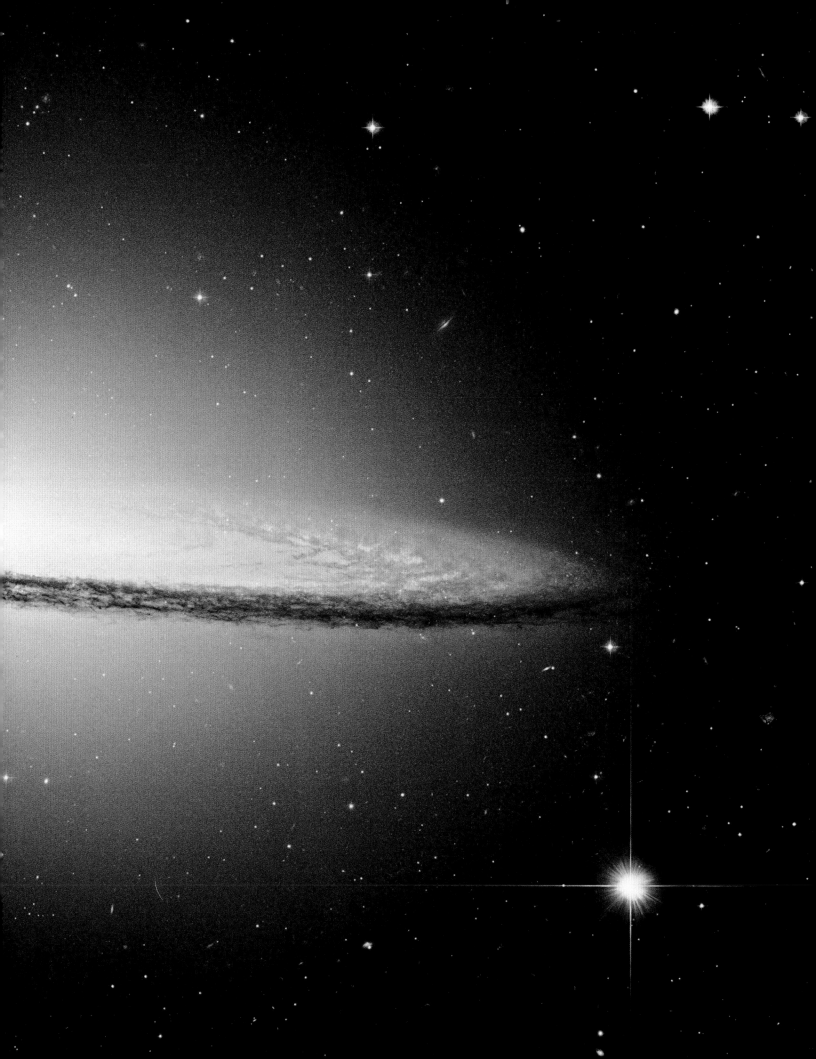

THE FIREWORKS GALAXY

Spitzer Space Telescope and Subaru Telescope
Infrared, Visible Light

- The Fireworks Galaxy, also known as NGC 6946, is one of the closest spiral galaxies to us, at a distance of about 10 million light-years. Found between the northern constellations Cepheus and Cygnus, it was discovered by William Herschel in 1798. NGC 6946 lies close to the band of the Milky Way in the sky, which passes through Cygnus. Thus, when we look at NGC 6946 we must peer through the stars and gas and dust in our own Milky Way, which is why so many foreground stars are visible in the images presented here.

  The image opposite shows NGC 6946 in visible light, taken by Japan's 8.2-metre Subaru Telescope on the summit of Mauna Kea in Hawaii. In addition to dark dust lanes and older stars, which appear orange, we can clearly see the presence of large red patches of light along the spiral arms. These are giant HII regions (see page 117), sites of active star formation where hot massive young stars are ionizing the surrounding hydrogen gas.

  The image below shows NGC 6946 in infrared light. The image was taken by the Spitzer Space Observatory and combines data at four different infrared wavelengths. Blue indicates emission at 3.6 microns, green shows emission at 4.5 microns, and red is from emission at 5.8 and 8.0 microns. The emission shown in red is from the dust which shows up as dark patches in the visible light image; this dust is very warm and is associated with star formation in the spiral arms. The blue dots spread across the image are from foreground stars in our own Milky Way.

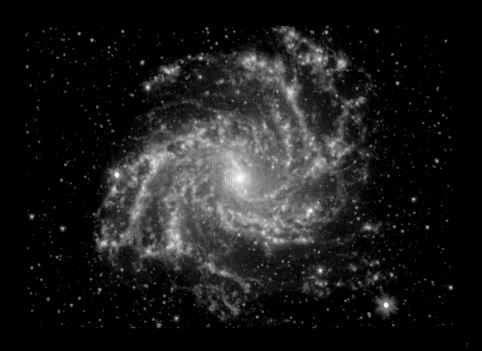

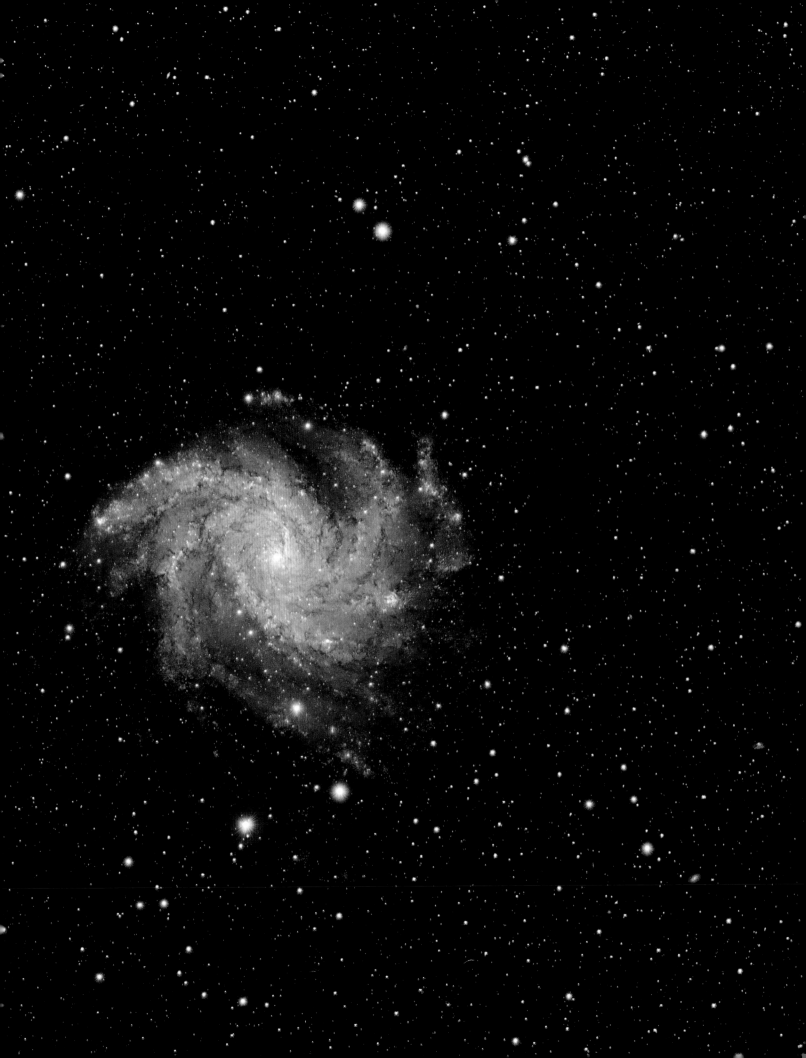

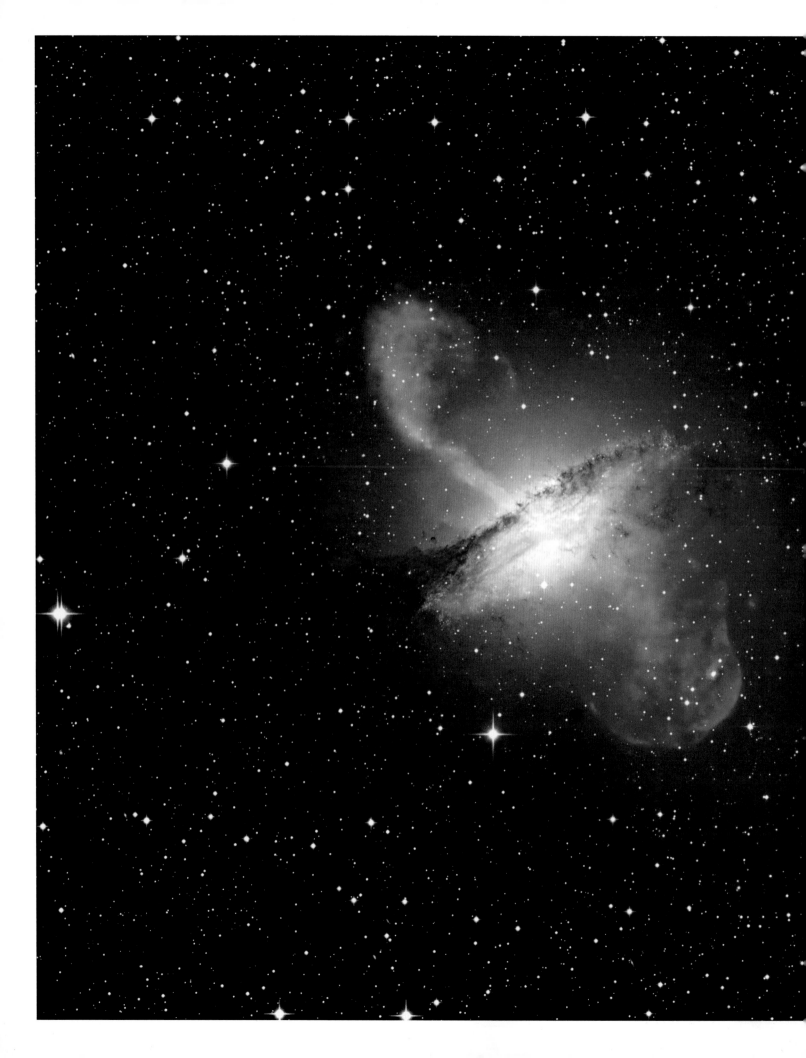

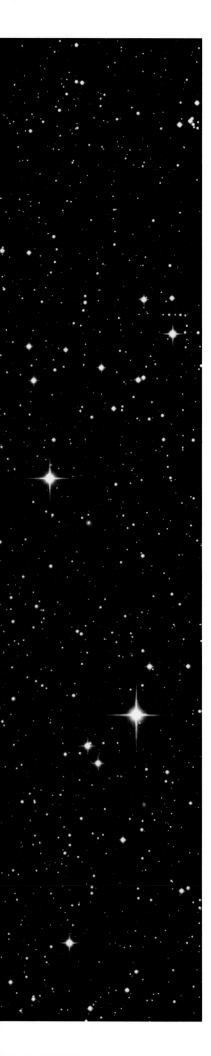

CENTAURUS A

Atacama Pathfinder Experiment, Max Planck Gezellschaft Telescope and Chandra X-ray Observatory
Submillimetre, Visible Light and X-ray

- Centaurus A gets its "A" designation from being the first radio source to be discovered in the southern constellation Centaurus in a radio survey of the sky in the late 1940s. The galaxy itself was first noted in 1826 by Scottish astronomer James Dunlop from his home in New South Wales, Australia. Due to its proximity and brightness it has become one of the most studied galaxies. It is also known as NGC 5128, and is the most prominent radio galaxy anywhere in the sky.

  The image shown here is a multi-wavelength image which combines submillimetre data (shown in orange), visible light data (shown in its true colours) and X-ray data (shown in blue). The visible appearance of Centaurus A has confounded astronomers for many centuries. Although much of the galaxy looks like a classical elliptical galaxy, the prominent dust lane suggests that it is a merger between two smaller galaxies.

  At Centaurus A's heart is a supermassive black hole; this ejects a jet of material moving at close to the speed of light at right angles to the dust lane. It can be seen in the millimetre and X-ray images, which extend far beyond the visible part of the galaxy. These giant lobes are what make Centaurus A such a bright galaxy at radio wavelengths. Radio observations of the inner parts of the jet show that it is moving at about half the speed of light, and that it extends over a million light-years from the centre. The X-rays are produced when the jet collides with surrounding gas, creating highly energetic particles.

NGC 891

Hubble Space Telescope
Visible Light

- NGC 891 is one of the closest examples of an edge-on spiral galaxy. It was discovered by William Herschel in 1784 using a six-inch telescope from his back garden in Bath. Herschel is more famous for being the first person to discover a new planet, Uranus, in 1781. NGC 891 lies in the northern constellation Andromeda, and is some 30 million light-years away. It is part of a group of galaxies named the NGC 1023 group, which is itself part of the Local Supercluster, also known as the Virgo Supercluster (VSC). The VSC is dominated by the Virgo Cluster, but also includes several smaller galaxy clusters, including our own Local Group, the Draco Group, the M81 Group, the Ursa Major Group and the M101 Group.

  The image shown here is taken by the Hubble Space Telescope's Advanced Camera for Surveys. The band of dust across the centre of the disk of the galaxy is clearly visible; this is how we believe our own Milky Way would look from a similar distance. However, if you look more closely you will also see filaments of dust extending above and below the dust lane, perpendicular to it.

  It is believed that these dust filaments are created by supernovae explosions in the disk of NGC 891 blowing the dust from the mid-plane of the disk. Stars form in the mid-plane of the disk where the concentration of gas is highest. Again, if you look closely at this image you can see the bluer stars in the mid plane of the disk; these are the hot young stars which will explode as supernovae after a few million years, a brief time in the life of the cosmos.

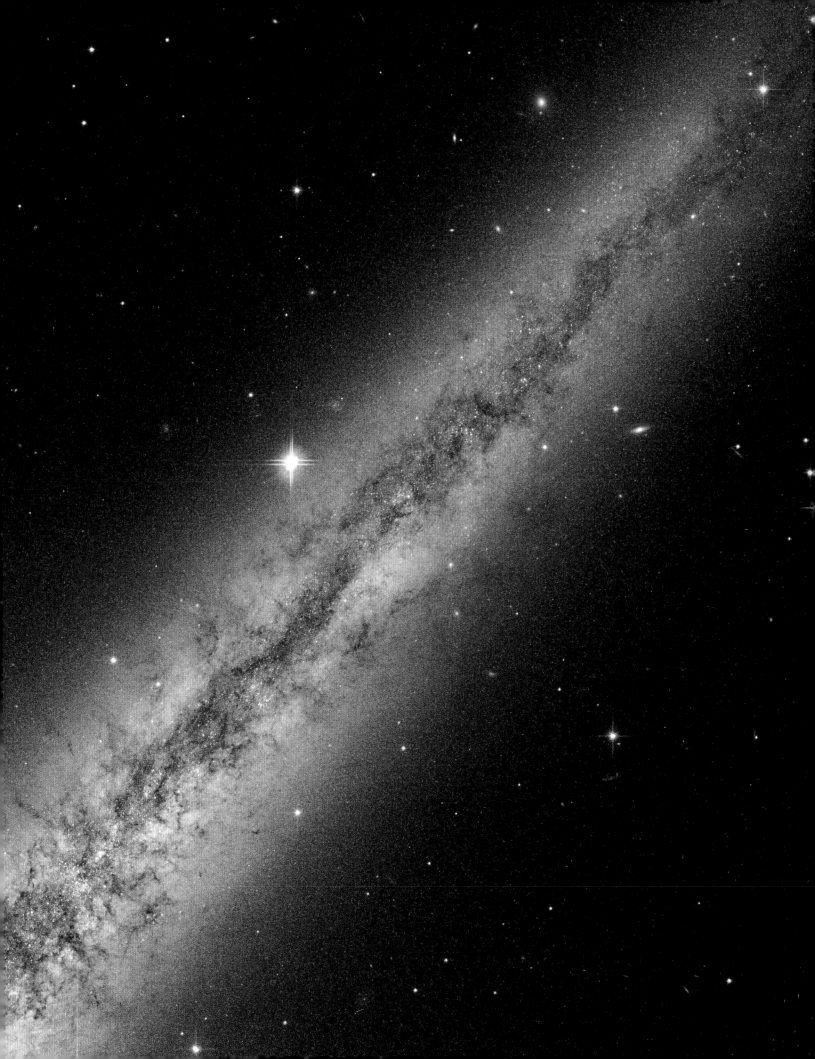

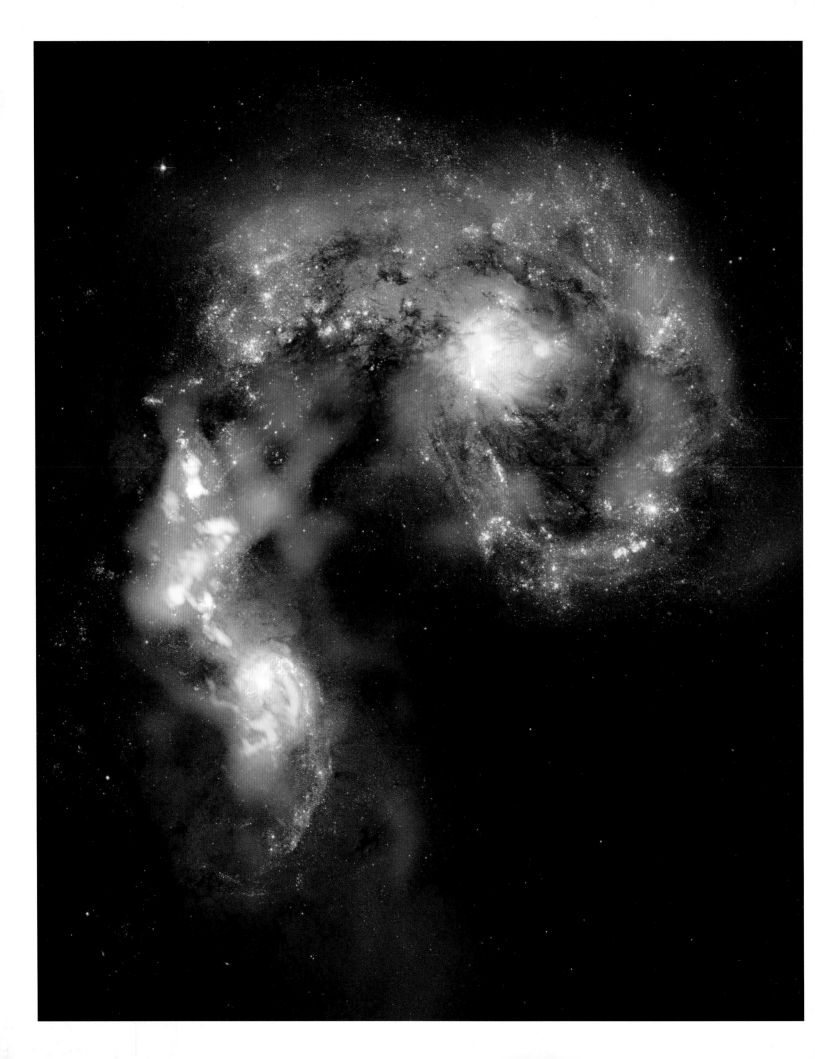

THE ANTENNAE GALAXIES

Atacama Large Millimeter Array and Hubble Space Telescope
Millimetre Waves and Visible Light

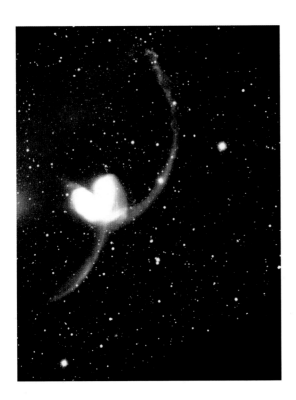

- The Antennae Galaxies are the closest and best studied example of two galaxies in the process of merging. The Antennae get their name from the two long tails of stars, gas and dust which can be seen in wide-field images of this merging system. At some 60 million light-years from Earth, there is no other pair of merging galaxies which allows us such a detailed view of what happens during such an event.

  It is believed that many galaxies, including the Milky Way, have undergone a merger event at some point in their history. Surprisingly, perhaps, the stars in the two merging galaxies do not collide, they are too small and there is too much space between them. But, the space between stars is filled with gas and dust, and when the gas in the two systems collides it triggers a huge burst of star formation.

  The main image is a composite view of the Antennae Galaxies. The background image is a visible light picture taken by the Hubble Space Telescope. Superimposed on this is a millimetre-wave image obtained by the Atacama Large Millimetre Array (ALMA). ALMA's millimetre and submillimetre views show up these extremely cold star-forming regions of space in red, pink and yellow.

  Located at an altitude of 5,000 metres in the Atacama Desert of northern Chile, ALMA is the largest and most expensive observatory ever built, an international collaboration between many countries including the USA, the countries of the European Southern Observatory, Canada, Japan and Chile.

Above: The two antennae-like tails are evident in this view from a ground-based telescope.

THE PINWHEEL GALAXY

Chandra X-ray Observatory, Galaxy Evolution Explorer, Spitzer and Hubble Space Telescopes
X-ray, Ultraviolet, Visible Light and Infrared

- The Pinwheel Galaxy, also known as Messier 101, is a face-on spiral galaxy in the constellation Ursa Major. It lies at a distance of about 21 million light-years from Earth, and with a diameter of about 170,000 light-years it is nearly twice the size of the Milky Way. It was discovered by Pierre Méchain in 1781 and added to Charles Messier's famous catalogue as one of its last entries. It was also extensively observed by Lord Rosse in the nineteenth century using his huge telescope, the Leviathan of Parsontown in Ireland.

  Although it has been observed and studied for centuries, the stunning image shown here is a very modern view of the Pinwheel Galaxy. It combines data from several of NASA's great observatories to reveal information that has never been seen before. In purple are data from the Chandra X-ray Observatory, showing emission from hot gas at millions of Kelvin. In blue are data in ultraviolet light from the Galaxy Evolution Explorer; here we see the emission from hot, massive young stars. As the image shows, these short-lived stars are found in the spiral arms of the galaxy where new star formation is taking place.

  Visible light data obtained with the Hubble Space Telescope are shown in yellow; this light is dominated by emission from stars like our Sun. These lower mass stars have much longer lifetimes, and so are found in the disk and bulge of the galaxy. Finally, data in infrared, taken by the Spitzer Space Telescope, is shown in red; here we see emission from the dust, which is also found in the spiral arms with the newly formed stars.

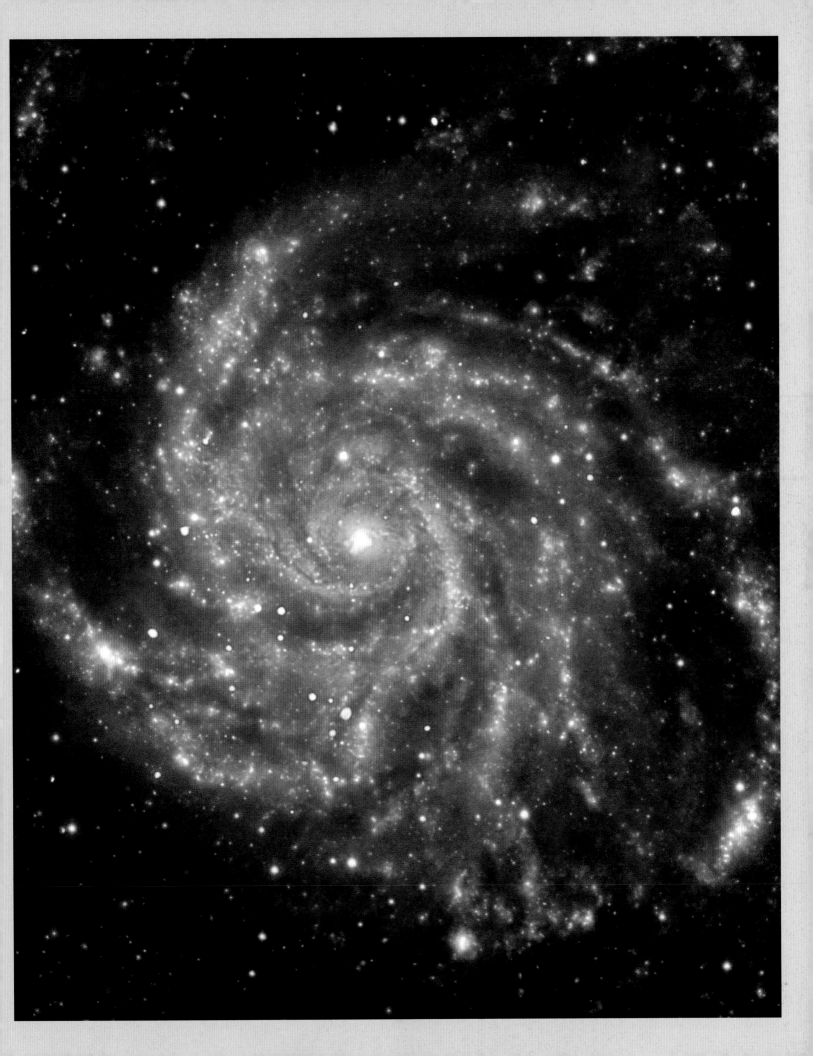

THE VIRGO CLUSTER

Mount Palomar Observatory's Schmidt Telescope
Visible Light

● The Milky Way Galaxy is part of the Virgo Supercluster of galaxies, which includes our Local Group and the Virgo Cluster. The Virgo Cluster includes the giant elliptical galaxy Messier 87 (NGC 4486), seen at centre of the image (left), and the two bright lenticular galaxies M86 and M84 above and to the right of M87. The Local Group is being drawn towards the Virgo Cluster by the huge mass of the cluster, which is estimated to be about one thousand trillion times the mass of the Sun.

The Virgo Cluster is approximately eight degrees across on the sky, which is some sixteen times the diameter of the Moon. Many of the brightest members of the Virgo Cluster were discovered in the late 1700s and early 1800s, and are easily observable through small telescopes. These include Messier 87 (NGC 4486) and its brightest member Messier 49, another giant elliptical galaxy. The main image was taken by the Mount Palomar Observatory's 48-inch Schmidt Telescope as part of the Palomar Sky Survey.

One of the key projects of the Hubble Space Telescope when it was launched in 1990 was to be able to see Cepheid variable stars out to the Virgo Cluster. Because of the period-luminosity relationship for Cepheids (see page 103), this would enable a more accurate value of the current rate of expansion of the Universe, the so-called Hubble constant, to be measured. The results of this key project were published in 2001, finding a value of 72 kilometres/second/megaparsec (+/– 8), achieving its target accuracy of a determination within an error of 10 percent. This value has since been confirmed by other techniques, including studies of the cosmic microwave background.

Below: This deep image of the Virgo Cluster was captured by the 60-centimetre Burrell Schmidt telescope. The dark circles are where bright foreground stars were burnt out on the original photographic plates. These have been removed to provide a better view of the galaxies in the cluster. Messier 87 is the giant elliptical galaxy at lower left.

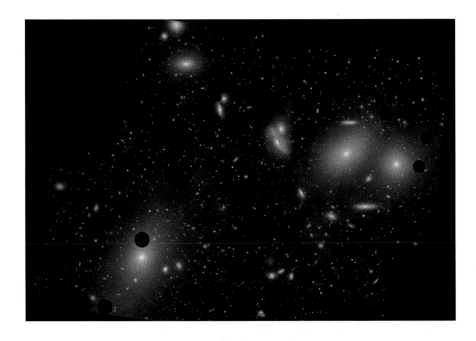

COLLIDING SPIRAL GALAXIES NGC 2207 AND IC 2163

Hubble Space Telescope
Visible Light

- NGC 2207 and IC 2163 are a pair of spiral galaxies which are colliding. Located in Canis Major, and lying about 80 million light-years from us, these spiral galaxies are at an earlier stage of merging than what we see with the Antennae Galaxies (see page 137). Currently they can still be considered two separate spiral galaxies; the tidal disruption which is normally so evident in merging galaxies is still at a fairly low level, as evidenced in this visible light image of the pair.

  However, closer inspection shows that NGC 2207, the larger of the pair (on the left), is in fact in the process of tidally stripping material from IC 2163. This tidal stripping has already led to an increase in massive star formation. Since 1975, four separate supernovae have been observed in this colliding pair, with three of

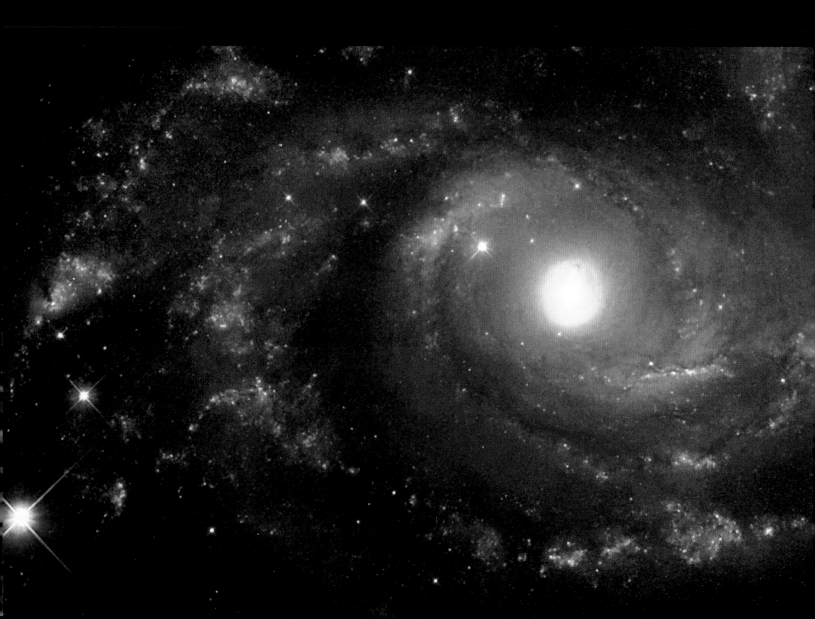

the four associated with the explosions of young, massive stars. As the collision leads to the merging of the two galaxies a much larger burst of star formation will occur throughout most of the system as the interstellar gas in the two spirals collides. This merger process is expected to take about one billion years, and when complete we speculate that either an elliptical galaxy or a disk galaxy will result.

NGC 2207 and IC 2163 were discovered in 1835 by John Herschel, William Herschel's son. This image is a real colour visible light image taken by NASA's Hubble Space Telescope.

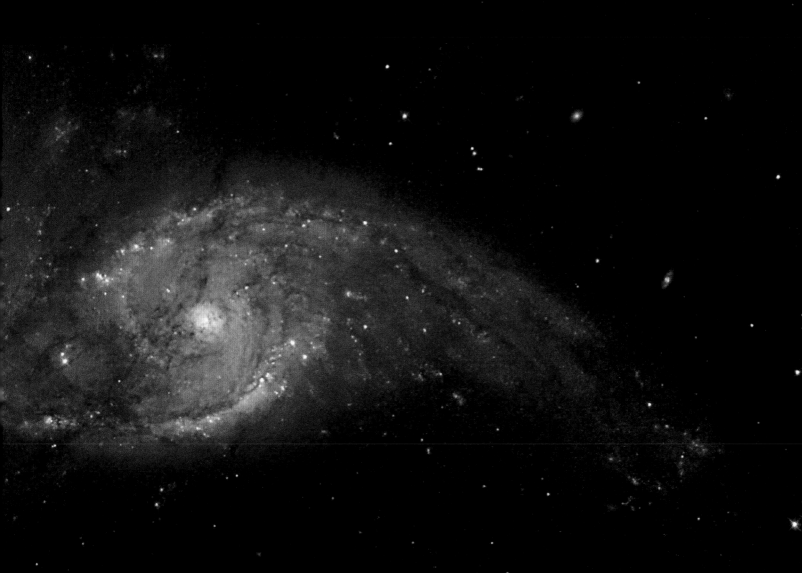

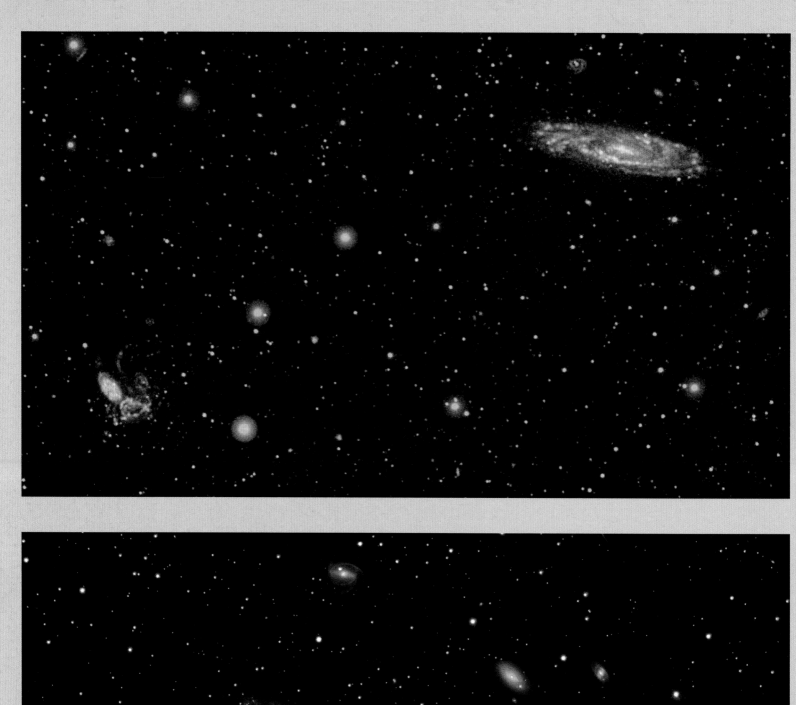

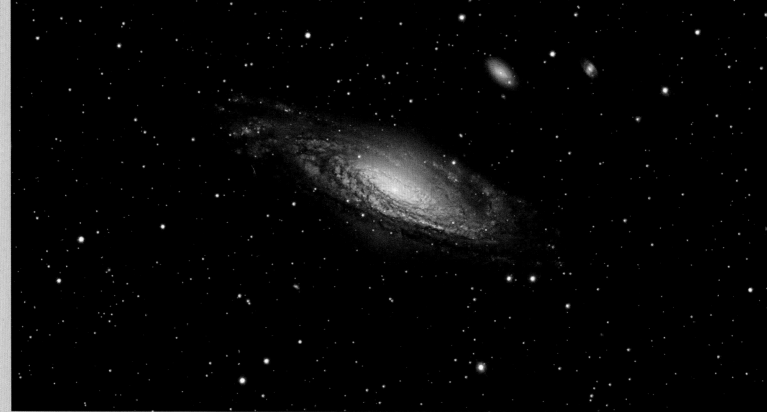

THE MILKY WAY'S TWIN, NGC 7331

Galaxy Evolution Explorer
Ultraviolet Light

• The spiral galaxy NGC 7331 is sometimes referred to as the Milky Way's twin, as it is believed to be similar in both size and structure. Located in the constellation Pegasus, and lying about 40 million light-years away, it was discovered by William Herschel in 1784. The picture left (top) was taken by NASA's Galaxy Evolution Explorer (GALEX), a satellite which imaged the sky in the ultraviolet (UV) part of the spectrum. NGC 7331 is in the upper right of the image, in the lower left is Stephan's Quintet (see page 149).

By imaging in UV light, GALEX is able to concentrate on the emission from the hottest, most massive stars in galaxies. Stars like our Sun are too cool to emit much UV light, but more massive stars burn hotter and their surfaces emit far more UV light than visible light. Such massive stars have much shorter lifetimes than stars like our Sun, and so when we image such massive stars we are also seeing the youngest stars in a galaxy.

The UV appearance of NGC 7331 clearly shows that these young, massive stars are only found in the spiral arms. Note the colour difference between the spiral arms and the bulge of the galaxy. The bulge contains lower mass, older stars, which emit much less UV light. Strong UV emission can also be seen in Stephan's Quintet: the tidal interaction between the four galaxies has triggered a massive burst of star formation, which pervades throughout the entire body of the galaxies, not just their spiral arms.

Opposite bottom: Visible light image of NGC 7331 acquired with the Schulman Telescope at the Mount Lemmon Skycenter Observatory, Arizona.

Below: The GALEX satellite being assembled in the clean room.

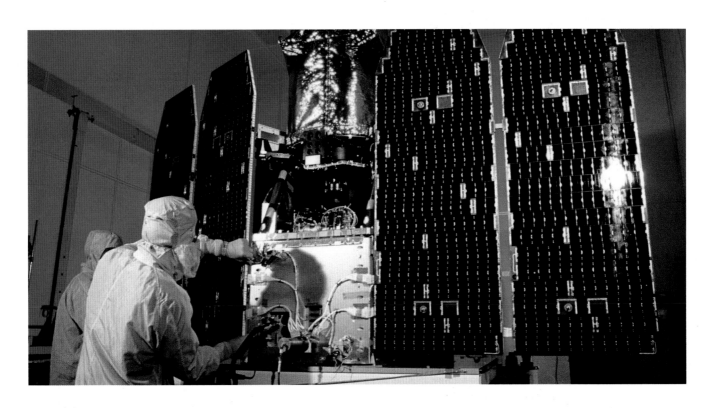

PANDORA'S CLUSTER

Chandra X-ray Observatory and Hubble
X-ray and Visible Light

- Some 80 percent of the matter in the Universe seems to be in the form of dark matter. This is matter which has a gravitational effect, but does not seem to interact with electromagnetic radiation. Dark matter was first suggested by Fritz Zwicky in the 1930s from studies of the Coma Cluster (see page 155). His suggestion was largely ignored, but by the 1980s new evidence had emerged to support the existence of dark matter, and now it has become a vital part in our understanding of the composition and evolution of the Universe.

  This is a composite image of Pandora's Cluster, also known as Abell 2744. The galaxies are shown in a visible light image taken by the Hubble Space Telescope. Superimposed on this is the X-ray emission (imaged by the Chandra X-ray Observatory), in red. The blue colour in the image is a calculation of the distribution of invisible dark matter in the cluster, based on studies of the motions of the galaxies in the cluster.

  Detailed studies of Pandora's Cluster have shown that galaxies make up less than five percent of the total mass of the cluster. Some twenty percent of the mass is in the form of hot gas between the galaxies; this gas is so hot that it only emits at X-ray wavelengths. Around 75 percent of the mass of this particular cluster is due to invisible dark matter, a form of matter which we are still trying to understand, although its gravitational effects are clearly visible in our studies of galaxies and clusters of galaxies.

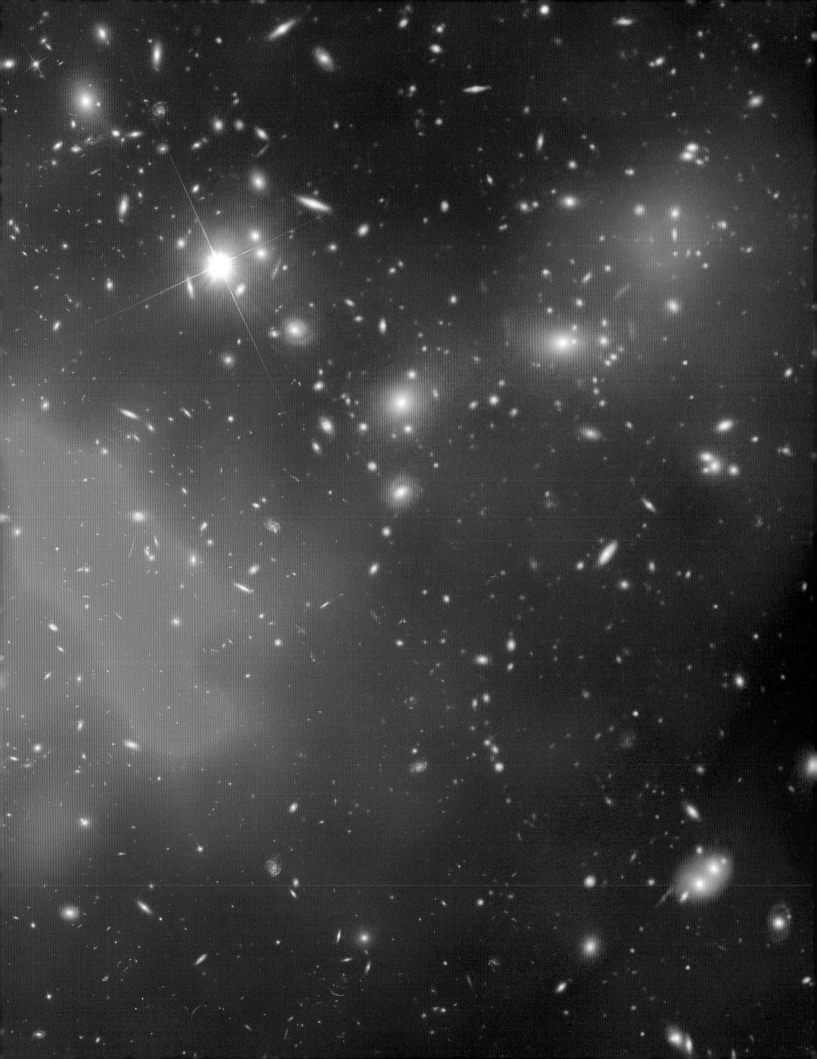

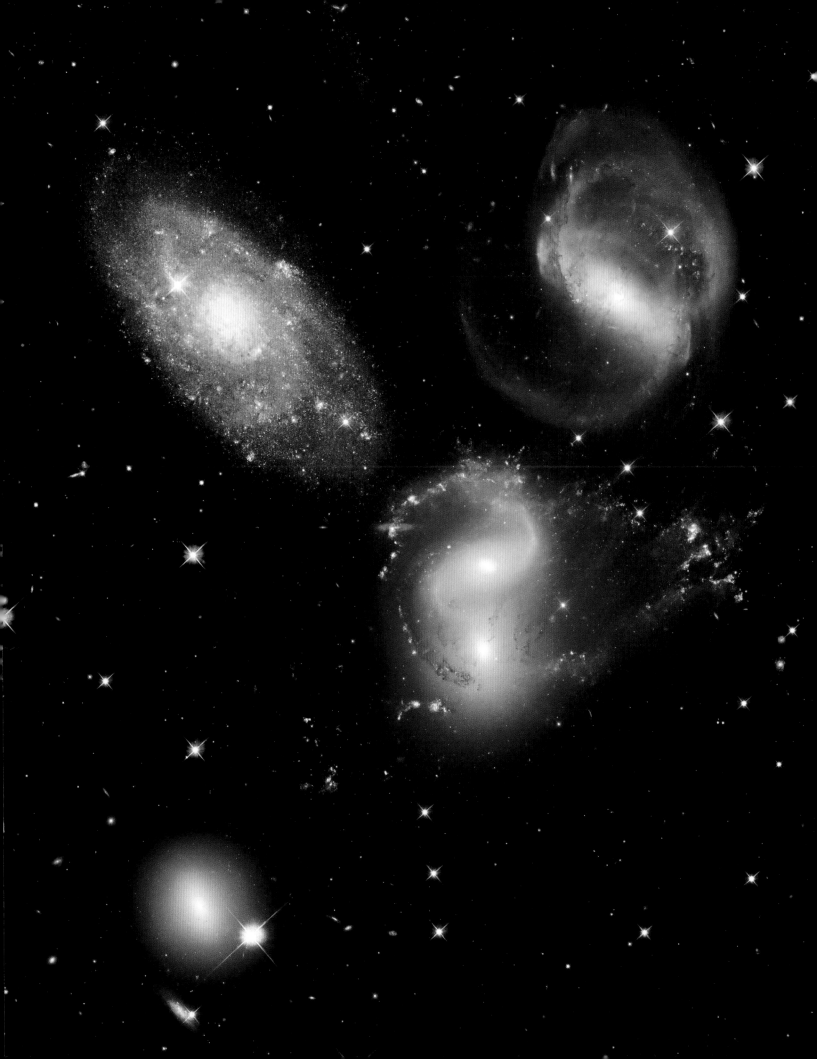

STEPHAN'S QUINTET

Hubble Space Telescope
Near-Infrared and Visible Light

- Discovered by Édouard Stephan at Marseille Observatory in 1877, Stephan's Quintet was the first compact group of galaxies ever found. One of the five galaxies in the quintet is just a visual alignment, but the other four actually form a physical association. They are involved in a cosmic dance, which will probably end with the four galaxies merging. So, in some ways, we are seeing an early version of the Antennae Galaxies, shown on page 137.

This image was taken in visible and near-infrared light using the Hubble Space Telescope and its Wide Field Camera 3 (WFC3), which was installed on the HST in May 2009. The image shows the quintet in their real colours, by combining images taken through near-infrared, red, green and blue filters. If you look closely at the galaxies you will notice that three of them have distorted shapes, including elongated spiral arms and tails. These distortions are due to the gravitational effects of the other galaxies; we call them tidal distortions. The interaction between the four galaxies in this compact group is leading to a flurry of star formation, particularly in the central pair of galaxies. This can be seen by the presence of very blue stars and large pink patches; these are HII regions which are sites of massive star formation being triggered by the tidal effects of the other galaxies. The near-infrared light is able to peer through some of the dust to see parts of the star formation which are hidden from view in visible light.

Studying Stephan's Quintet allows us to witness the evolution of a compact system of galaxies. We are seeing this system change from being a group which would have been dominated by the emission of visible light from spiral galaxies into a more developed system typical of the centres of rich clusters of galaxies. In these galaxies the emission is dominated by visible light from elliptical galaxies and X-ray emission from hot gas between the galaxies.

OVERLAPPING GALAXIES

Hubble Space Telescope
Visible Light

- This pair of overlapping galaxies is known as NGC 3314. At first glance they look as if they are merging, but in fact they are a chance alignment and are not physically associated with each other. How do we know this? The most obvious clue is that there is no sign of the tidal disruption which we see when galaxies gravitationally interact and merge. Secondly, such a merger would trigger a burst of star formation, and no such burst is seen in this pair. Thirdly, we can measure the redshift of each galaxy and from this calculate their distances. This shows that NGC 3314a is in the foreground at a distance of some 120 million light-years, whereas the background galaxy NGC 3314b is a further 20 million light-years behind it.

  Further clues are evident if we look carefully at the details of the background galaxy, NGC 3314b. The dust lanes in the background galaxy appear less pronounced than those in the inner parts of the foreground galaxy NGC 3314a. This is due to the foreground galaxy reducing the starlight from the background galaxy, due to its own dust content. By reducing and reddening the starlight coming through the foreground galaxy, the contrast of dust lanes in the background galaxy are reduced.

  This striking image was taken by the Hubble Space Telescope using its visible light Advanced Camera for Surveys (ACS), which has been on the telescope since March 2002. The ACS replaced the last original instrument, the Faint Object Camera.

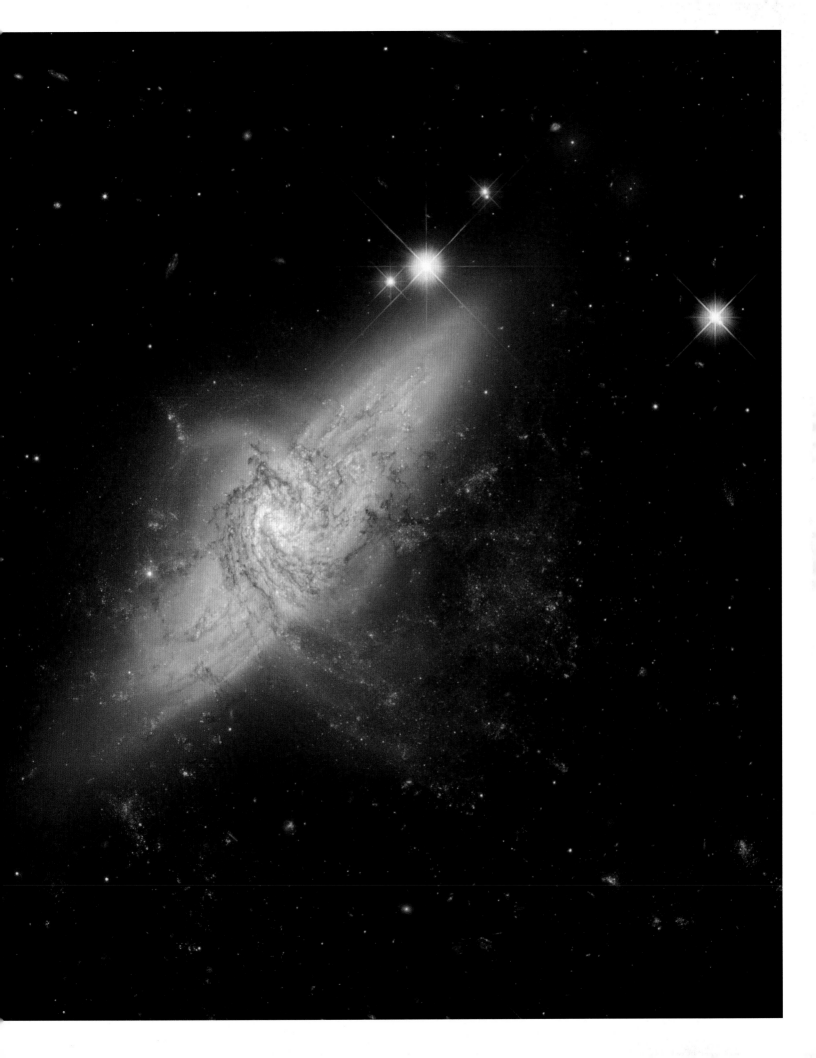

BARRED SPIRAL GALAXY NGC 1433

Hubble Space Telescope and Atacama Large Millimetre Array
Visible, Millimetre Waves

- Many spiral galaxies are found to contain a bar across their inner parts. In fact, approximately two thirds of spiral galaxies are found to contain a bar, including our Milky Way Galaxy. The presence of a bar can affect the motion of stars and gas in the galaxy, it can affect the activity of the nucleus of the galaxy, and it can affect the structure of the arms, too.

  The origins of bars is still being debated, but it is thought that they are temporary phenomena in the lives of spiral galaxies, with the bar structure decaying over time and leading to a normal spiral galaxy. However, given how common they are, it is also thought that they are a recurring phenomenon, they decay and then regenerate in a cyclical fashion during the lifetime of a galaxy.

  The galaxy shown below, NGC 1433, is one of the closest examples to us of a barred spiral galaxy, lying at a distance of some 30 million light-years. The image combines visible light data from the HST (shown in blue to represent shorter wavelengths) with a millimetre wavelength image taken by the Atacama Large Millimetre Array (ALMA) shown in colour. The ALMA image traces the presence of gas, dust and stars which are in the process of being formed. It reveals for the first time a spiral structure near the galaxy's nucleus, as well as outflows of gas and dust, due to the vigorous star formation occurring deep in the nucleus of this active galaxy. The image opposite shows a visible light image taken by the Hubble Telescope; the bar is clearly visible and has a prominent dust lane across it.

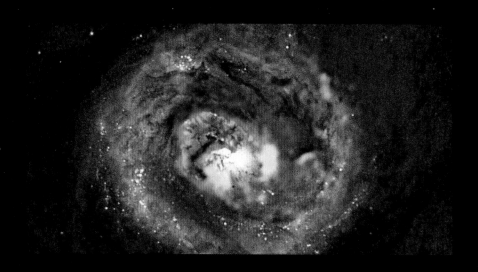

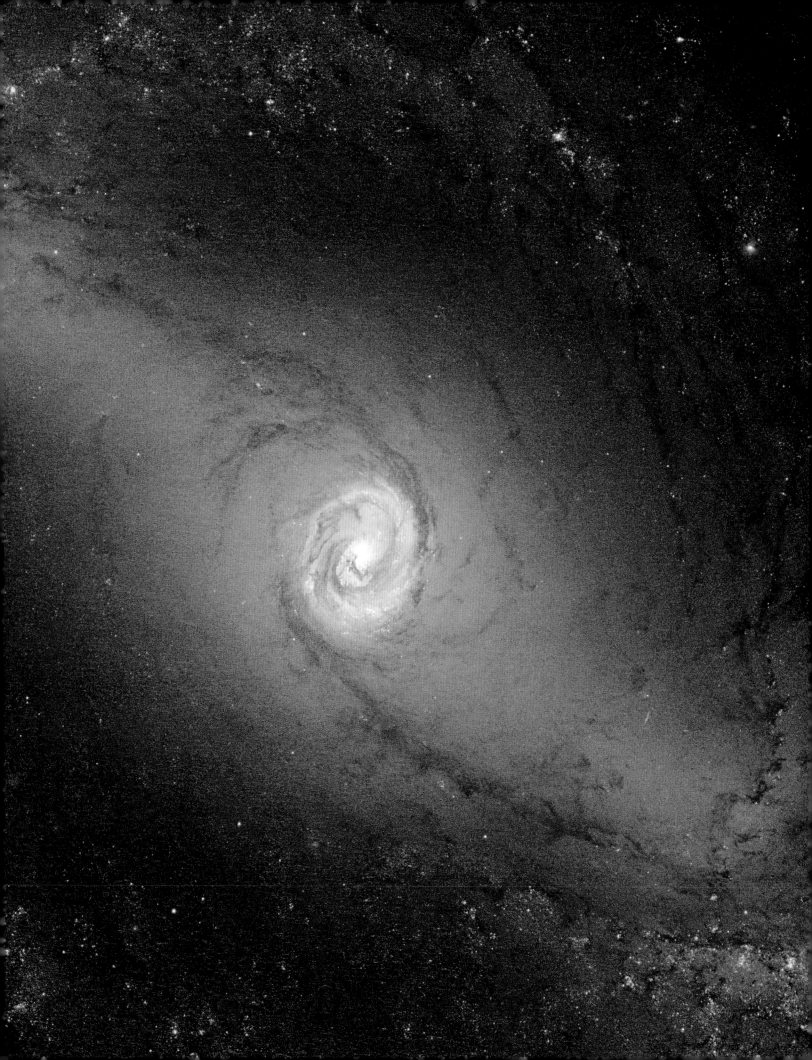

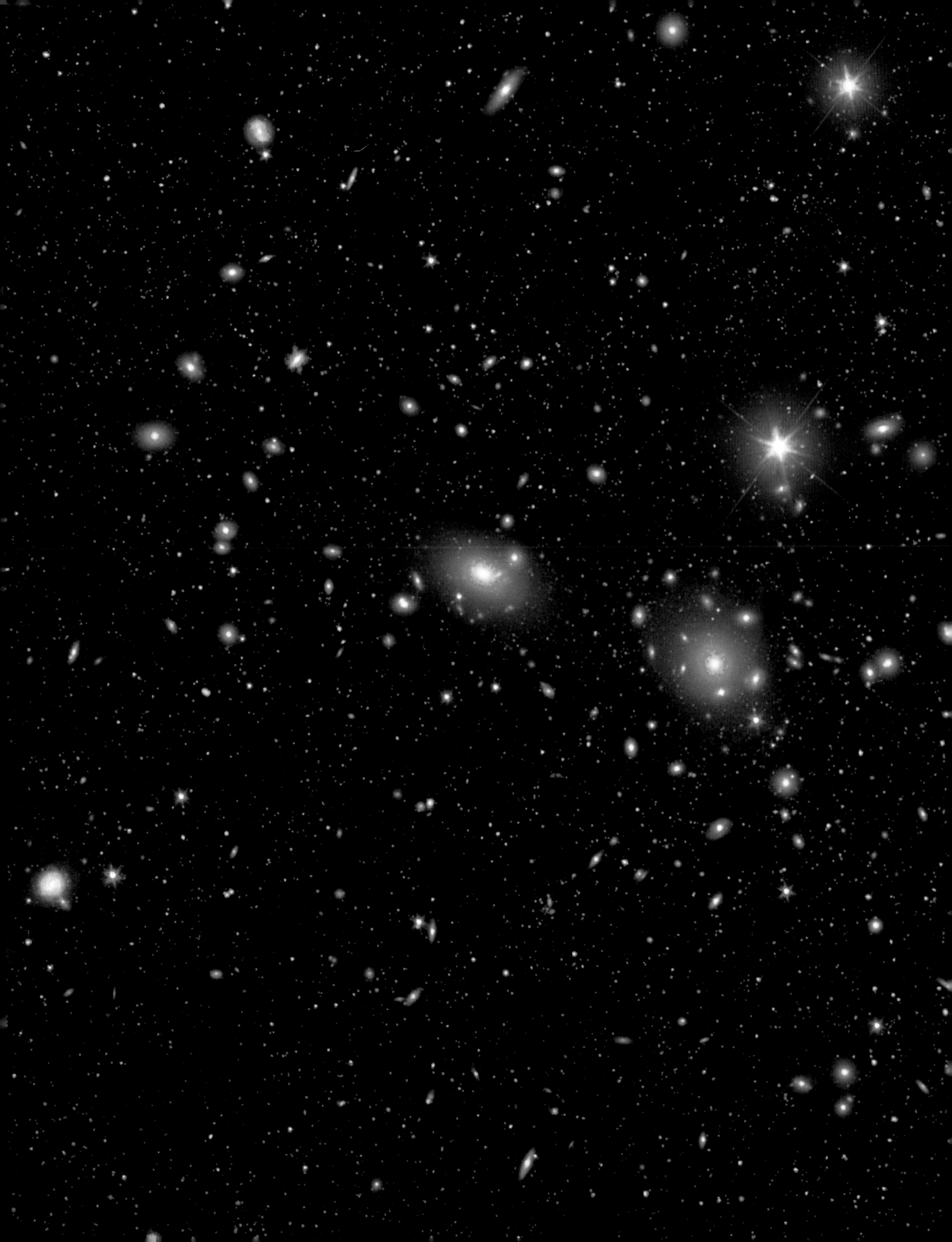

THE COMA CLUSTER

Hubble Space Telescope and Sloan Digital Sky Survey
Near-Infrared and Visible Light

- The Coma Cluster is a relatively close cluster of galaxies containing more than one thousand identified members. It lies at a distance of about 330 million light-years from us and, along with the Leo Cluster, it is a major component of the Coma Supercluster. The cluster contains about a dozen spiral galaxies which are bright enough to be observed by telescopes with a 20-centimetre aperture, making this cluster a favourite among amateur astronomers. As is the case with most rich clusters of galaxies, nearly all of the galaxies in this cluster are elliptical and lenticular galaxies, and its central regions are dominated by two giant elliptical galaxies NGC 4874 and NGC 4889.

  In 1933 the Swiss astronomer Fritz Zwicky did a detailed study of the motions of the galaxies in the Coma Cluster using the Mount Wilson 100-inch Telescope. By looking at the redshifts of the galaxies he came to the conclusion that the galaxies were moving too quickly for the cluster to be held together by the gravitational forces of the visible material. He suggested instead that the cluster must be held together by what he called "dunkle Materie", what we now call dark matter. His work was ignored until the 1980s, when separate evidence for dark matter was discovered.

  This is a composite image combining near-infrared data from the Hubble Space Telescope with shorter wavelength near-infrared and visible light data from the Sloan Digital Sky Survey. Almost every object you see in this image is a galaxy, each containing billions of stars.

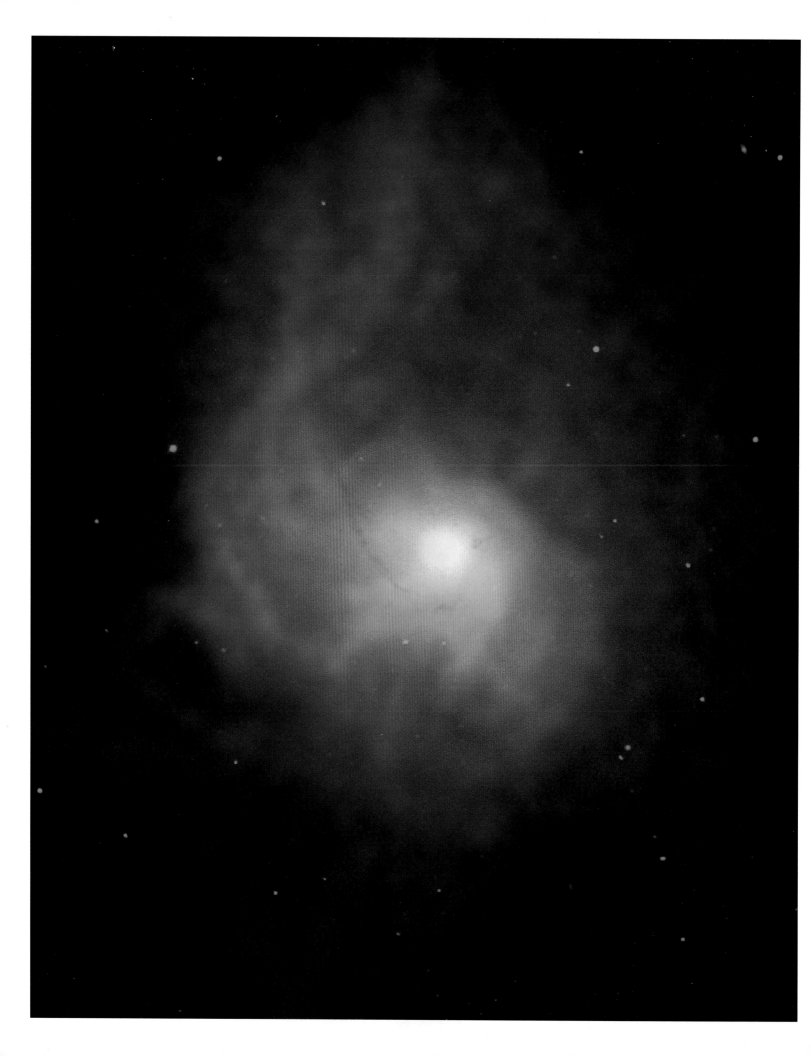

CORE OF THE CENTAURUS CLUSTER IN X-RAYS

Very Large Array, Spitzer Space Telescope and Chandra X-ray Observatory
Radio Waves, Infrared Light and X-rays

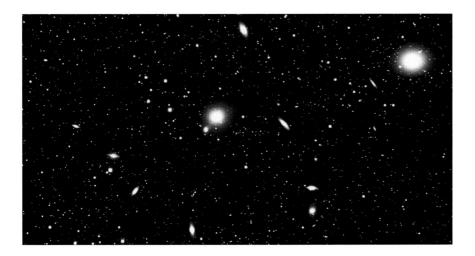

- At the centre of the Centaurus Cluster (also known as Abell 3526) lies the giant elliptical galaxy NGC 4696. This composite image of NGC 4696 shows a vast cloud of hot gas (red), surrounding high-energy bubbles 10,000 light-years across (blue) on either side of the bright white area around a supermassive black hole. The green dots are infrared radiation from star clusters at the edge of the galaxy.

  The Centaurus Cluster consists of hundreds of galaxies, many of which are shown in the accompanying visible light image taken by the 4-metre Blanco Telescope at the Cerro Tololo Observatory in Chile.

  The Centaurus Cluster is located about 170 million light-years away in the Centaurus constellation, and is part of the Hydra-Centaurus supercluster, along with the IC 4329 Cluster and the Hydra Cluster. The X-rays are emitted by hot gas at millions of Kelvin which lies between the galaxies in the cluster – this gas is known as the intracluster medium. The gas attains these very high temperatures by the release of gravitational energy during the formation of the cluster from smaller structures.

  Until X-ray observations of clusters became available, this hot intracluster gas was not known to exist. We now know that most of the normal (baryonic) mass of a cluster is in this intracluster medium, about 80–85 percent, rather than in the stars that we see in visible light. However, the baryonic mass of clusters is itself just a minority fraction of the total mass. Measurements show that about 20 percent of the mass of clusters seems to be in the form of baryonic matter; the remaining 80 percent is believed to be in the form of dark matter, the nature of which we are still trying to understand.

Above: The Centaurus Cluster (Abell 3526) imaged in visible light by the 4-metre telescope at Cerro Tololo.

THE FLYING V, IC 2184

Hubble Space Telescope
Infrared and Visible Light

- IC 2184, also known as the Flying V, shown opposite top, is actually two distinct objects. This Hubble Space Telescope visible and infrared image shows this interacting pair of galaxies; collectively they are known as IC 2184. We see both galaxies nearly edge-on (left), and stretching out into space we can see the tidal tails being formed by the interaction of the two galaxies. These tidal tails are streams of gas, dust and stars which are disrupted out of the body of each galaxy by the gravitational force of the other galaxy.

  IC 2184 is located in the constellation Camelopardalis, a faint constellation in the northern part of the sky. The galaxies lie about 165 million light-years from us, and were first noted by French astronomer Guillaume Bigourdan in 1900. Observations suggest that the two galaxies are barred spiral galaxies; the tidal tails which are normally curved appear straight because we are seeing the system edge-on.

  Also visible in the Hubble image are areas of bright blue. These are bursts of recent star formation, areas where the gases from the two interacting galaxies are colliding and triggering the creation of copious numbers of new stars. We are seeing IC 2184 at the early stages of their merger: as the two galaxies continue to merge they will eventually form a single, larger galaxy, with intense star formation similar to what we currently see in the Antennae Galaxies (see page 137). What the nature of the ensuing merged galaxy will be is still the subject of intense debate.

Opposite bottom: This Hubble image of merged galaxies NGC 6240 shows that the resulting shape after a merger of galaxies can be highly irregular.

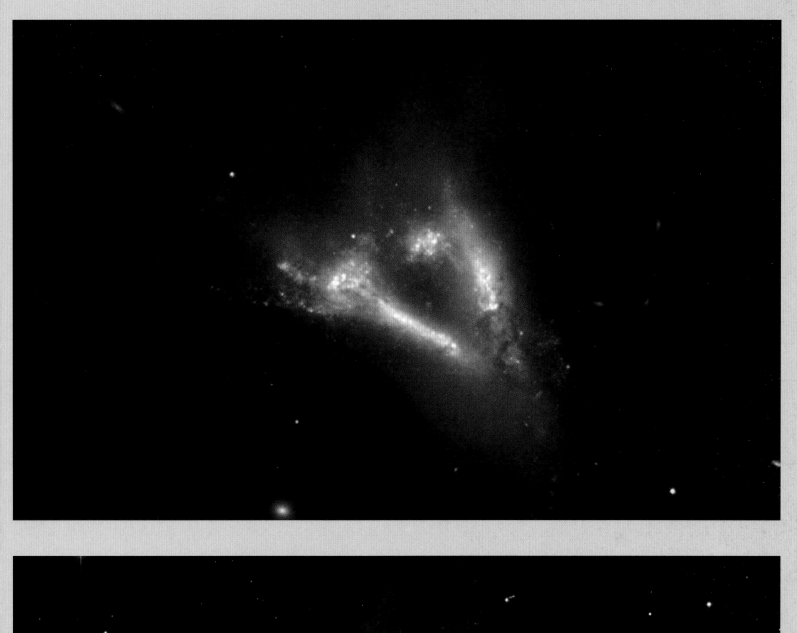

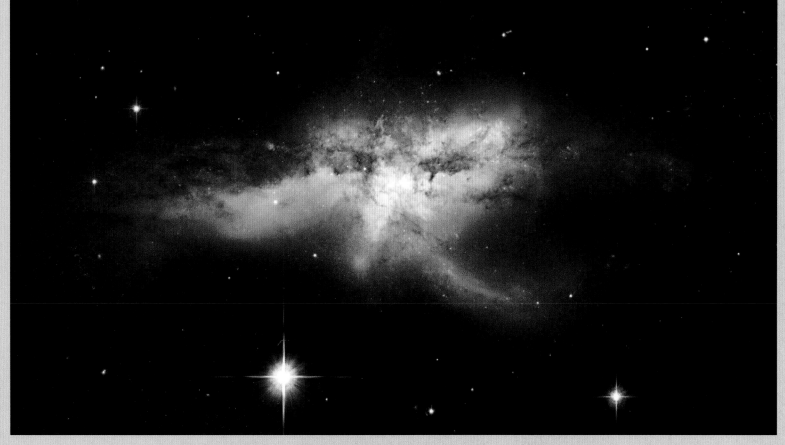

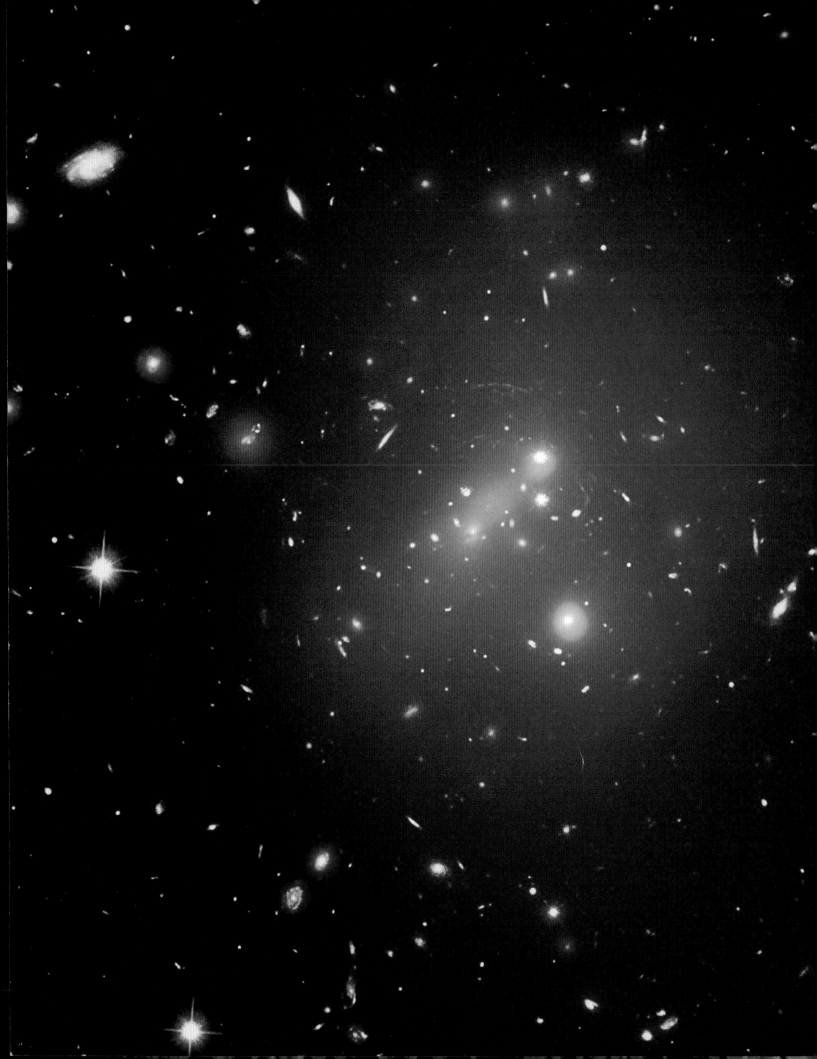

THE MOST DISTANT CLUSTER

Spitzer and Hubble Space Telescopes and Chandra X-ray Observatory
Infrared, Visible Light and X-rays

- IDCS 1426 is the most massive galaxy cluster detected at such a young age. It was first discovered by the Spitzer Space Telescope in 2012. The main image shows a combination X-ray image taken by the Chandra X-ray Observatory, visible light image taken by the Hubble Space Telescope (HST), and infrared-light taken by the Spitzer Space Telescope. The X-ray emission is shown in blue, the visible light in green, and the infrared light in red. The image below shows the visible light image taken by the HST on its own, without the other wavelengths.

After its discovery, the distance to IDCS 1426 was determined by measuring its redshift. Using the HST and the 10-metre Keck Observatory on the summit of Mauna Kea, IDCS 1426 was found to be at an incredible distance of about 10 billion light-years. This means that it can only be a few billion years old, and it is the most massive galaxy cluster ever seen at such an early age. Measurements suggest that about 90 percent of its mass is in the form of dark matter.

The brightest X-ray emission comes from near the centre of the cluster, but not at its centre. This suggests that the cluster has already had a collision or interaction with another massive system of galaxies. This is not surprising; it is believed that such massive structures would only have had enough time to form in the early Universe if they resulted from the merger of smaller clusters. It looks like we are seeing IDCS 1426 after such a merger.

Below: A visible light image of IDCS 1426
from the Hubble Space Telescope.

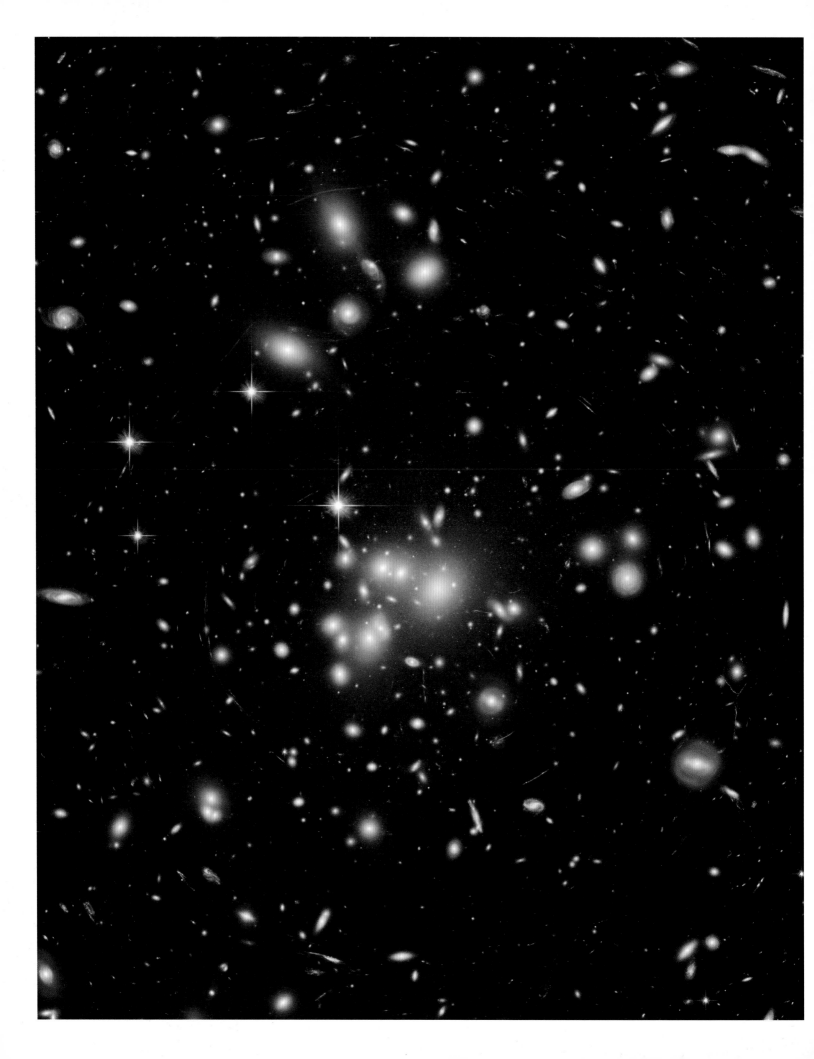

ABELL 1689

Hubble Space Telescope and Chandra X-ray Observatory
Infrared, Visible Light and X-rays

- Abell 1689 is one of the brightest and most massive galaxy clusters discovered. Lying in the constellation Virgo, at a distance of just over two billion light-years from us, its enormous mass leads to it acting as a gravitational lens (see page 179), distorting images of galaxies which lie behind it. It also has more than 160,000 globular clusters, the largest population ever found in a galaxy cluster.

  The main image is a combined visible light and infrared image taken by the Hubble Space Telescope (HST). The combined exposure time of this image is over 34 hours. The galaxies in the cluster can be seen as orange objects. There are, however, many background galaxies which have been gravitationally lensed by Abell 1689; these show as red and blue in this image. Some of them can also be distinguished by their curved shape, forming a small arc of a circle which is centred on the cluster's centre. When it was discovered in 2008, one of the lensed galaxies, A16890zD1, was the most distant galaxy found. This record has since been surpassed, for example by the galaxy GN-z11 shown on page 174.

  The image below shows the same visible light and infrared image from the HST combined with an X-ray image taken by the Chandra Space Observatory. The X-ray emission is due to hot intracluster gas emitting at millions of degrees. The large number of gravitationally lensed background galaxies in this system has allowed us to study the amount and the distribution of dark matter in Abell 1689 in detail.

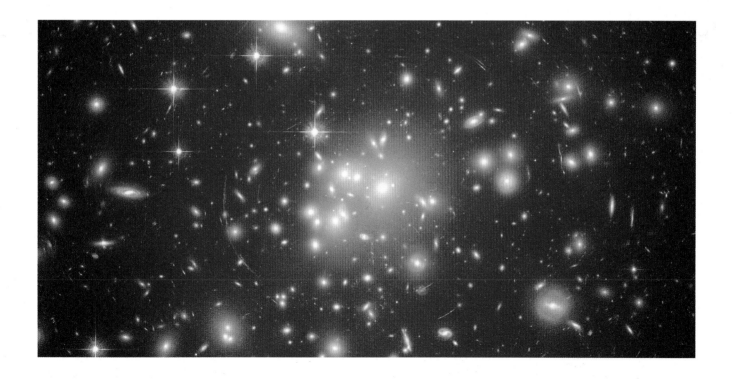

THE BULLET CLUSTER

Hubble Space Telescope and Chandra X-ray Observatory
Infrared, Visible Light and X-rays

- The best evidence for dark matter is said by many to be provided by the Bullet Cluster. The combined image (right) shows this evidence; the visible light image was taken by the Hubble Space Telescope, the X-ray image (shown in red) was obtained by the Chandra X-ray Observatory, and the blue shows the distribution of mass in the system as determined by the gravitational lensing of background galaxies.

  The Bullet Cluster is, in fact, two colliding clusters of galaxies. This collision has led to the Bullet Cluster being used to argue for the existence of dark matter. In the collision of two clusters, the stars in the galaxies are largely undisturbed. The space between stars is so large compared to their sizes that they pass through each other when galaxies and clusters collide. The gas between the stars and between the galaxies behaves differently, colliding and being heated to high temperatures by the collision. At these high temperatures the gas emits X-rays, and this emission is shown in red. The visible light seen by HST and the X-ray light seen by Chandra trace the distribution of normal, baryonic matter in the Bullet Cluster.

  The gravitational lensing of background galaxies provides us with a way of tracing the distribution of all mass in the colliding clusters, whether normal baryonic matter or not. When we do this we obtain the distribution shown in blue; this quite clearly is different from the distribution of the baryonic matter. It is this difference which, it is argued, provides such strong evidence for the reality of dark matter in this system.

Below: The Musket Ball Cluster is an example of the aftermath of a collision between two galaxies, which has resulted in the separation between dark and normal matter being observed. This newly discovered system has been nicknamed the Musket Ball Cluster, because it is older and slower than the Bullet Cluster. The location of the majority of the matter in the cluster, dominated by dark matter, is coloured blue. The hot gas is coloured red. Where red and blue overlap, the result is purple.

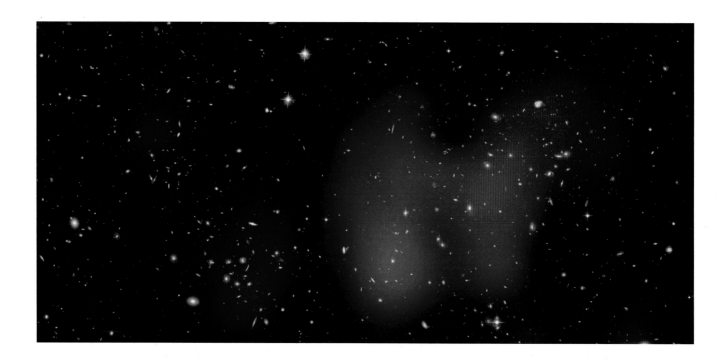

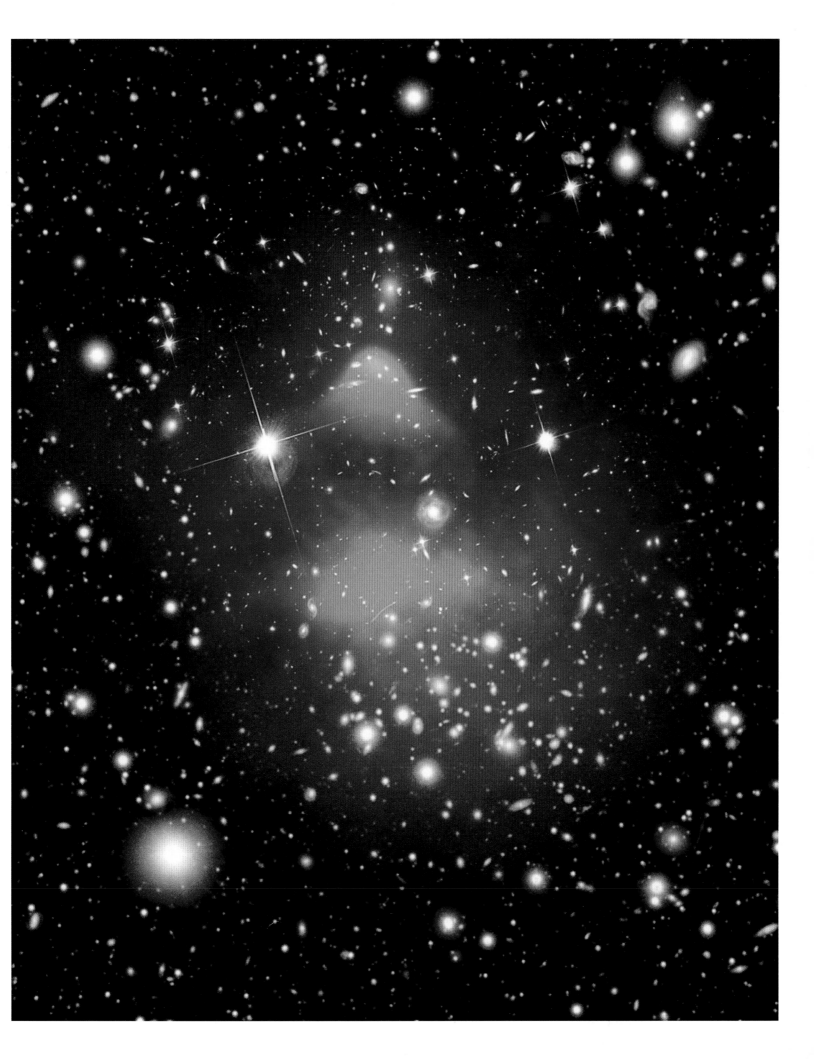

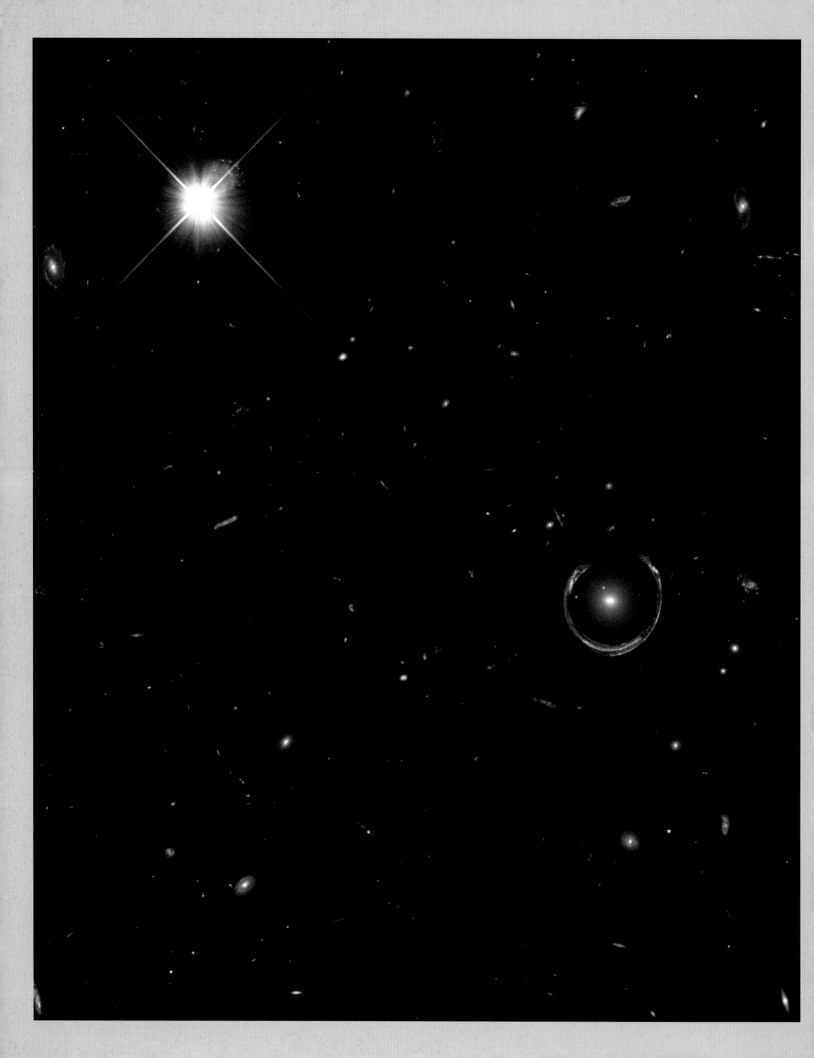

HORSESHOE EINSTEIN RING

Hubble Space Telescope
Infrared and Visible Light

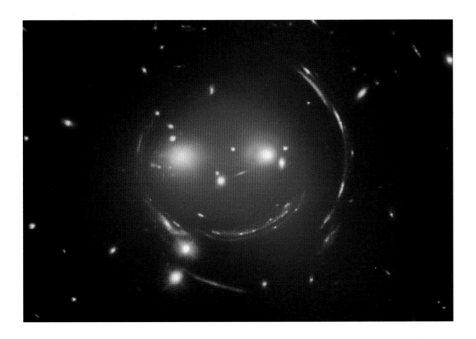

- The horseshoe shape in this incredible image (left) is, in fact, a galaxy. It is a background galaxy being gravitationally lensed by a foreground galaxy, the luminous red galaxy at the centre of the horseshoe. Luminous red galaxies are a type of galaxy that contain about 10 times the mass of the Milky Way, and their large mass makes them ideal candidates for producing gravitational lenses. But, luck must also play a part. For a gravitational lens to occur, the massive galaxy must lie between us and a more distant galaxy, as it does here.

This particular gravitational lens is known less romantically as LRG 3-757, and was discovered by the Sloan Digital Sky Survey (SDSS) in 2007. The image shown here is a Hubble Space Telescope image, and combines visible light and infrared. Most gravitational lenses produce two distinct images of the background galaxy. However, in this particular case the alignment of the foreground and background galaxies is so precise that the two images of the background galaxy have been distorted into an almost complete ring around the foreground galaxy.

Not only are gravitationally lensed galaxies beautiful to look at, they are also useful scientifically. They allow us to measure the total gravity of the foreground lensing object, teaching us about dark matter. The gravitationally lensed galaxy is often magnified, allowing us to see faint distant galaxies which we would otherwise not be able to detect. The background galaxy here is over 10 billion light-years from us, but appears brighter than it otherwise would.

Above: Another famous gravitationally lensed ring is named the Cheshire Cat, seen here in a combined visible light image from the Hubble Space Telescope and an X-ray image from the Chandra X-ray Observatory. Each "eye" galaxy is the brightest member of its own group of galaxies, and these two groups are moving together at over 500,000 kilometres per hour. Data from NASA's Chandra X-ray Observatory (purple) show hot gas that has been heated to millions of degrees, which is evidence that the galaxy groups are slamming into one another.

PART 5

AT THE EDGE OF
THE UNIVERSE

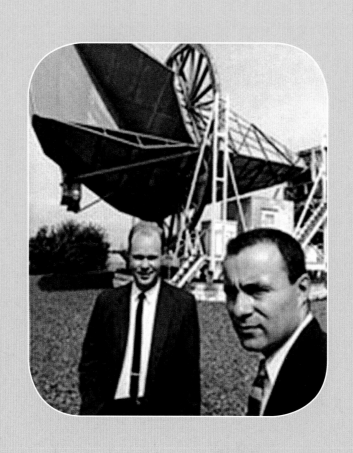

As we look further into space, we also look further back in time. Although light travels extremely fast, at 300 thousand kilometres every second, the vast distances involved in astronomy mean that the light from any object in space takes a finite time to get to us. In the Solar System these times are brief; only eight minutes for the light from the Sun and some 80 minutes for the light reflected from Saturn. But, the light-travel times quickly become appreciable.

The light from the next nearest star to us, Proxima Centauri, takes just over four years to reach us, and we are seeing many of the stars in the night-time sky as they were hundreds of years ago. The light from our nearest neighbour, the Andromeda Galaxy, has taken more than two million years to reach us, it started on its journey long before any modern humans existed. Light from the Virgo Cluster has taken more than 50 million years to travel to us, and light from more distant objects has taken hundreds or even thousands of millions of years. Light from the most distant objects we can see, at the very edge of the observable Universe, has taken more than 13 billion years to reach our telescopes. Our modern telescopes are able to see back to galaxies as they were when the Universe was just a few hundred million years old.

In the early 1960s the most popular model of the Universe was the Steady State theory. This argued that the Universe had always existed, and had not changed over time. The competing theory, the Big Bang theory, was believed by only a small minority of physicists. This all changed in 1965 with the accidental discovery of a relic radiation from the early Universe, a radiation which we now call the cosmic microwave background. Other discoveries, such as that of quasars, led to the realization that our Universe has indeed changed over time, contrary to the predictions of the Steady State theory. Detailed studies of the cosmic microwave background have allowed us to measure the age of the Universe, its expansion rate, and even its composition.

Large surveys of the positions and distances of hundreds of thousands of galaxies have shown us that our Universe is made up of vast sheets of galaxies called superclusters, with enormous empty spaces called voids separating them. In February 2016 we observed gravitational waves for the first time from two merging black holes. This has opened up a whole new window on the Universe, providing us with a method of studying the Universe in an entirely new way. It is hoped that gravitational waves will allow us to see back to the very earliest fraction of a second after the Big Bang, a time which is hidden from us using any kind of electromagnetic radiation.

Opposite: Robert Wilson (left) and Arno Penzias standing in front of the horn-reflector antenna on Crawford Hill in Holmdel Township, New Jersey, which detected the microwave background radiation and corroborated the Big Bang theory.

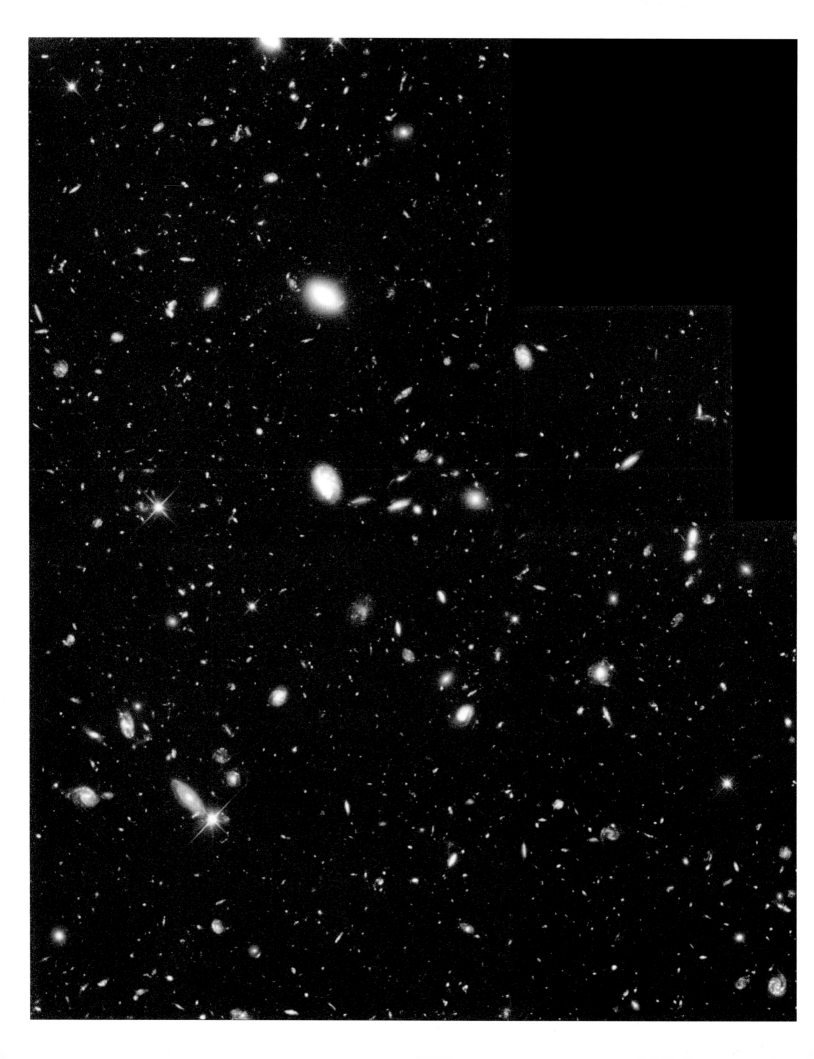

HUBBLE DEEP FIELD

Hubble Space Telescope
Visible Light

- In December 1995 the Hubble Space Telescope stared at an apparently blank part of the sky in the constellation Ursa Major for 10 consecutive days. In total, the Wide Field and Planetary Camera 2 (WFPC2) made 342 separate exposures of the same patch of sky, an area only 2.5 arc-minutes across (for comparison, the full Moon is 30 arc-minutes across). The image that resulted from this 10-day exposure has gone on to become one of the most important and iconic images taken by Hubble to date. Rather than seeing nothing in this "blank" part of the sky, instead thousands of galaxies were revealed. Nearly everything you see in the image at left is a distant galaxy. Because such a small patch of the sky was imaged, very few foreground stars are present, and these can easily be distinguished by their characteristic diffraction pattern (the streaks of light emanating from some of the stars). Instead, the Hubble Deep Field (HDF) shows nearly 3,000 galaxies, many of them very young and very distant.

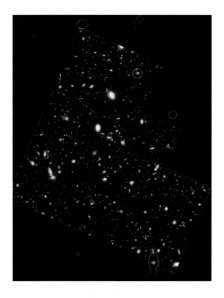

The Hubble Deep Field helped redefine our understanding of the early Universe. It showed that galaxies formed at a much earlier epoch in the Universe's history than many had previously thought. The Hubble Deep Field has also become one of the most cited images in the history of astrophysical research; by the end of 2015 it had been cited in about one thousand different astronomical research papers.

Multiwavelength observations of the Hubble Deep Field followed its release in 1995. In 1998 two important follow-ups were published, observations in the sub-millimetre part of the spectrum and observations in the radio part of the spectrum. The 850-micron sub-millimetre image (below right) was obtained using the Sub-millimetre Common User Bolometer Array (SCUBA) camera on the James Clerk Maxwell Telescope (JCMT). The JCMT is a 15-metre telescope dedicated to sub-millimetre astronomy on the summit of Mauna Kea in Hawaii. At an altitude of 4,200 kilometres, the dry air at the summit allows this radiation, which does not penetrate the atmosphere to sea level, to be detected.

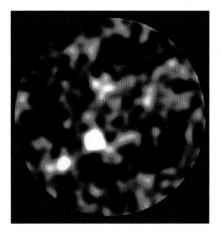

The radio image shown (top right) was obtained using the Very Large Array in New Mexico. It was taken at a wavelength of 3.5 centimetres and it detected seven radio sources in the HDF, all of which correspond to galaxies that can be seen in the original visible light image. The SCUBA image, although at a much lower resolution, initially detected five sources. All of these sub-millimetre sources correspond to extremely distant galaxies with a distance of more than seven billion light-years; this is because this part of the spectrum is detecting dust and is an indicator of massive star formation. Distant galaxies show much higher rates of star formation than nearby ones, so it is these distant galaxies with a high star-formation rate which dominate the sub-millimetre image.

Top: Radio data from the Very Large Array (shown as yellow contours) superimposed on the Hubble Space Telescope's visible light image of the Hubble Deep Field.

Above: James Clerk Maxwell Telescope sub-millimetre view of the Hubble Deep Field.

HUBBLE DEEP FIELD IN INFRARED

Spitzer Space Telescope
Infrared

- Distant galaxies with a high redshift do not emit much visible light for several reasons. Firstly, any visible light which they emit will have been redshifted into the infrared part of the spectrum. Secondly, it might be expected that early galaxies would be enshrouded in obscuring dust due to intense star formation. Therefore, soon after the Hubble Deep Field (HDF) was obtained, observations in the infrared were undertaken of the same patch of the sky.

In the late 1990s the Infrared Space Observatory (ISO) indicated that the infrared emission from 13 of the galaxies visible in the HDF far exceeded their visible light output, suggesting the presence of copious quantities of dust associated with intense bursts of star formation. These observations were followed up by the more sensitive Spitzer Space Telescope, which was launched in 2003.

The image (right top) shows the result of Spitzer's infrared observation of the HDF. Using the visible light image obtained by the Hubble Space Telescope the nearer galaxies were removed; these are the grey patches you can see in the Spitzer image (right bottom). What remains are very distant infrared-bright galaxies, some of them emitting hundreds of times more light in the infrared than they are in visible light. These are some of the earliest galaxies so far observed in the evolution of the Universe.

Below: Taken by the Near-Infrared Camera and Multi-Object Spectrometer (NICMOS) aboard NASA's Hubble Space Telescope, the image is part of the Hubble Ultra Deep Field survey, a deeper portrait of the Universe.

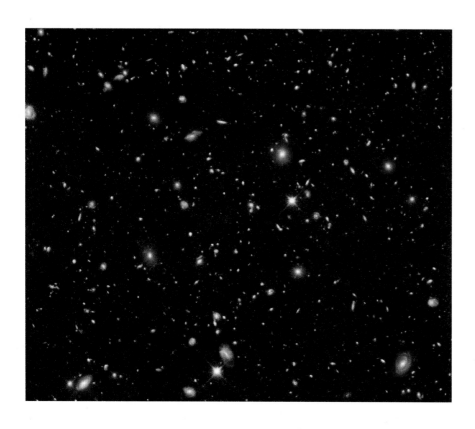

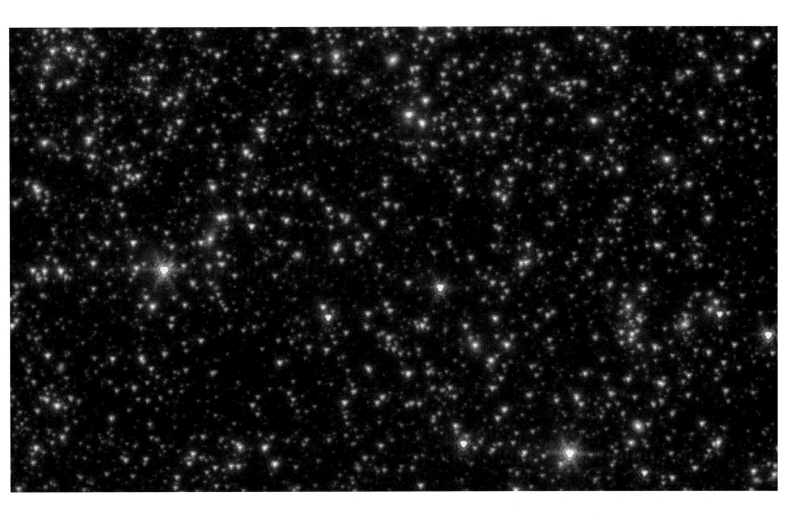

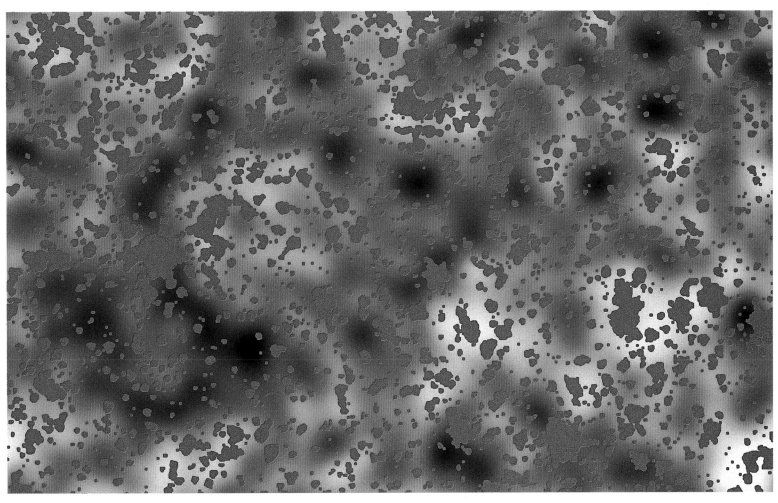

THE MOST DISTANT GALAXY EVER SEEN

Hubble Space Telescope
Infrared, Visible and Ultraviolet Light

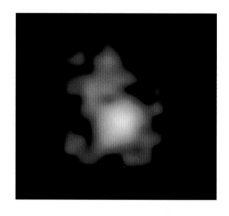

- In March 2016 it was announced that the Hubble Space Telescope (HST) had spotted the most distant galaxy ever detected (right). Named GN-z11, the galaxy was imaged in the infrared by HST's Wide Field Camera 3 (WFC3). Its strange appearance led to astronomers deciding to investigate it further. Its spectrum was obtained, also by HST's WFC3, in order to measure its redshift and hence infer its distance. GN-z11 was found to have a redshift of $z=11.1$, which means that the light from this galaxy has taken an incredible 13.4 billion years to reach us.

  We are seeing GN-z11 as it was when the Universe was only 400 million years old. It is believed that the first stars did not form until the Universe was about 100 to 200 million years old, so we are witnessing this galaxy just a little later than this. GN-z11 is estimated to have a mass of about a billion times the mass of the Sun, and is producing stars at a prodigious rate.

  This discovery has pushed the HST to its very limits. As we look back to earlier and earlier times, the light from distant galaxies is increasingly shifted into the infrared part of the spectrum by the expansion of the Universe. In October 2018 NASA will launch the James Webb Space Telescope. This 6.5-metre telescope will operate from the red part of the visible part of the spectrum (wavelengths longer than 0.6 microns) and into the infrared, out to wavelengths of 27 microns. One of its primary goals is to look for the most distant objects in the Universe, so we expect this current record of $z=11.1$ to be beaten in the not too distant future.

Above: The most distant galaxy yet discovered, GN-z11, and at right the location of GN-z11 in Hubble's GOODS North survey.

Below: An artist's impression of how the James Webb Telescope will look when complete.

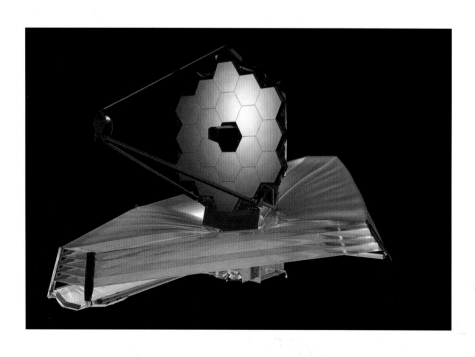

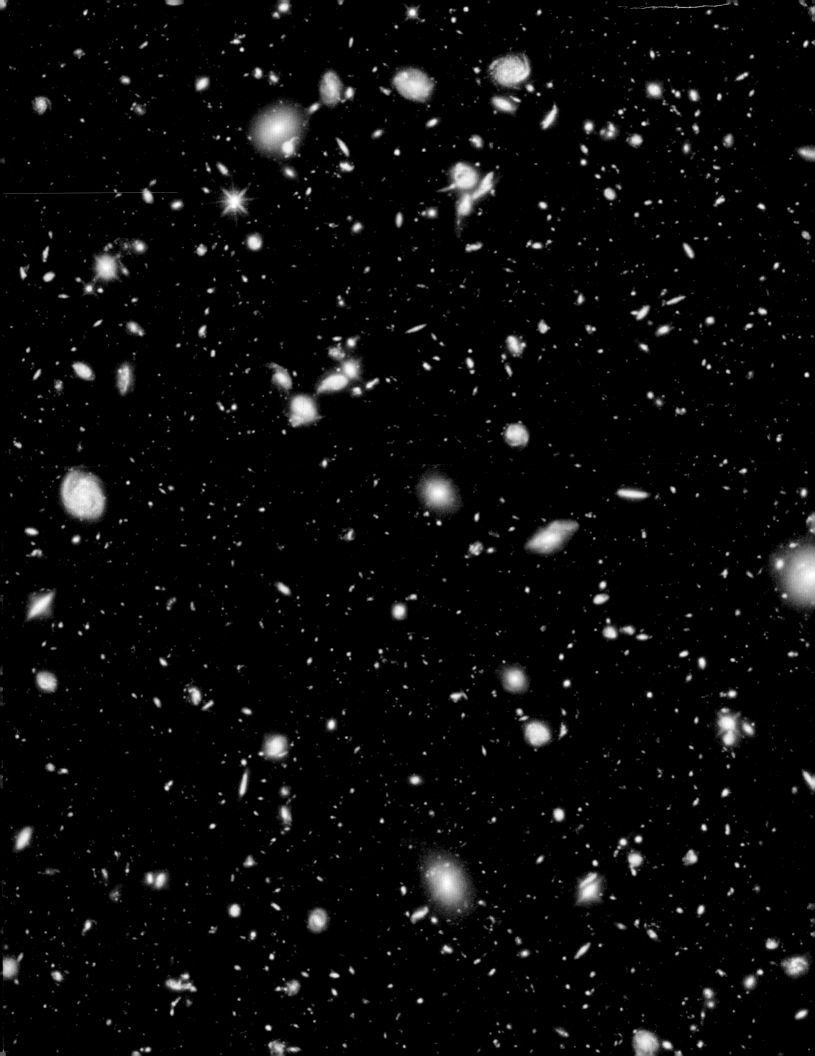

THE HUBBLE EXTREME DEEP FIELD

Hubble Space Telescope
Infrared and Visible Light

- Building on the success of the Hubble Deep Field (see page 171), three years later in 1998 NASA used the same camera, the Wide Field Planetary Camera 2 (WFPC2), to obtain a similarly deep image of part of the southern sky, called Hubble Deep Field South. In March 2002 WFPC2 was replaced by the Advanced Camera for Surveys (ACS). With a higher sensitivity than WFPC2, it was realized that the ACS could obtain a deeper image of a small patch of the sky than the Hubble Deep Field (HDF).

  It was decided to devote 400 orbits of the Hubble Space Telescope (HST) to obtain this new image. Taken between September 2003 and January 2004, the ACS imaged a small rectangular patch of sky measuring 2.4 by 2 arc-minutes in the Fornax constellation. The ACS's entire wavelength range was used, from the ultraviolet to the near-infrared. This image is known as the Hubble Ultra-Deep Field (HUDF), and contains an estimated 10,000 galaxies seen back to about 13 billion years ago.

  In September 2012 NASA released an even deeper image of the central part of the HUDF, combining images taken over the previous 10 years. The total exposure time was two million seconds, or just over 23 days. Known as the Hubble eXtreme Deep Field (XDF), this image covers 2.3 by 2 arc-minutes, approximately 80 percent of the area of the HUDF. It is the deepest visible light view of space that we currently have. XDF adds another 5,500 galaxies to the 10,000 seen by the HUDF and shows galaxies as far back as 13.2 billion years ago. The faintest galaxies visible in the XDF are ten-billionths of the brightness the human eye can see; many of the smallest galaxies seen will evolve into larger galaxies like the Milky Way by merging with other galaxies.

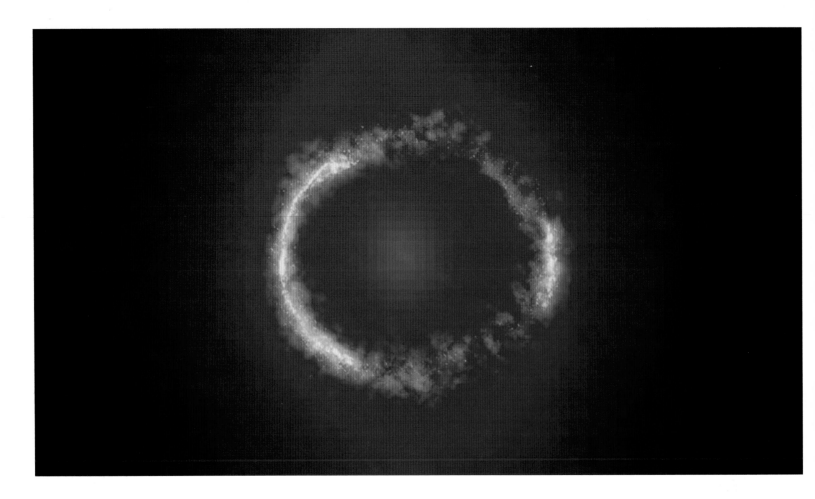

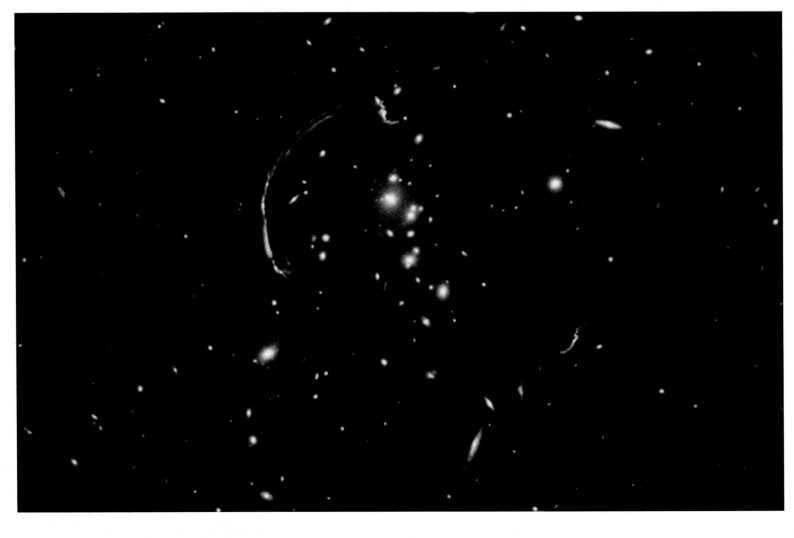

A GRAVITATIONAL LENS, SDP81

Atacama Large Millimetre Array and Hubble Space Telescope
Millimetre-Wave Infrared and Visible Light

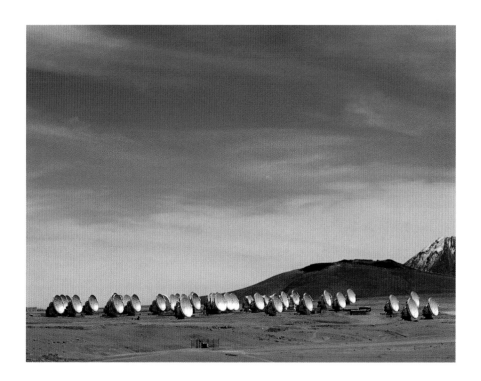

- In a paper written in 1936 Albert Einstein acknowledged the theoretical possibility of foreground objects gravitationally lensing more distant objects, but dismissed the possibility of ever observing such a phenomenon. He was, however, wrong. In 1988 the first ever gravitationally lensed object was discovered when radio astronomers saw a background quasar being lensed by a foreground galaxy, creating two separate images of the background quasar.

Since then, hundreds of similar gravitational lenses have been discovered. The one shown opposite is known as SDP81 and was discovered in 2010 by the Herschel Space Observatory, the largest infrared satellite ever launched. The top left image was obtained by the Atacama Large Millimetre Array (ALMA) in October 2014. Below it is a visible light image of the same object, taken with the Hubble Space Telescope.

SDP81 itself is about 12 billion light-years away, and is an actively star-forming galaxy. The foreground galaxy which is lensing it is much closer, about four billion light-years away. Not only does the lensing create an arc, one of the most tell-tale signs of gravitational lensing, but it also brightens the background galaxy just as a normal lens can brighten a faint source. The ALMA image is a high-resolution millimetre wavelength image of the dust emission from the primitive background galaxy being lensed by the intervening galaxy.

Above: ALMA, the Atacama Large Millimeter/submillimeter Array. It is the largest millimetre–submillimetre telescope in the world. It stands on the Chajnantor plateau in northern Chile at an altitude of 5,000 metres. When completed, it will consist of 66 12-metre and 7-metre antennae, which are moveable and can be located up to 16 kilometres apart.

THE EARLY UNIVERSE

Cosmic Background Explorer, Wilkinson Microwave Anisotropy Probe and the Planck Satellite
Microwave Radiation

- In April 1992 NASA's COsmic Background Explorer (COBE) satellite found evidence for the seeds of the structure in our Universe. By studying tiny temperature variations in its earliest light, the cosmic microwave background (CMB), COBE was able to produce the image shown top right. The CMB was produced when the Universe was about 400,000 years old; before then the Universe was opaque, so the CMB is the earliest light we can see. Due to the expansion of the Universe since then, the CMB's light has been red-shifted from the visible into the microwave part of the spectrum.

   Observing the microwave part of the spectrum from the ground is very difficult; for the same reason that we use microwave ovens to heat our food, water absorbs microwaves very effectively. COBE was launched in 1989 to be able to study the CMB properly without the atmosphere's water vapour impeding our view. In early 1990 COBE showed that the CMB's spectrum was a perfect blackbody at a temperature of 2.725 Kelvin, an important part of the argument that it is due to radiation from a hot, early Universe.

   It took two more years of observations to get the necessary data to produce this image, which shows tiny variations about the average temperature of 2.725 K, and by tiny we mean tiny – less than ten-thousandths of a Kelvin. These tiny variations correspond to areas of higher or lower density in the early Universe, and these are the seeds of the clusters and superclusters of galaxies that we see in the Universe today.

   Since COBE first detected the anisotropies in the cosmic microwave background (CMB) in 1992, two other satellites have been launched to study the CMB in more detail. In 2001 NASA launched the Wilkinson Microwave Anisotropy Probe (WMAP), Wilkinson in honour of Dave Wilkinson, one of the pioneers of CMB research. In 2009 ESA launched Planck, the first European satellite designed to study the CMB.

   Both WMAP (middle right) and Planck (bottom right) have returned increasingly detailed images of the anisotropies in the CMB. Studying these detailed images has allowed cosmologists to accurately determine many of the Universe's most important parameters such as its age, its expansion rate, its geometry and its composition. The last 25 years has seen us move into an era of "precision cosmology", being able to determine precise details about the Universe from studying the CMB.

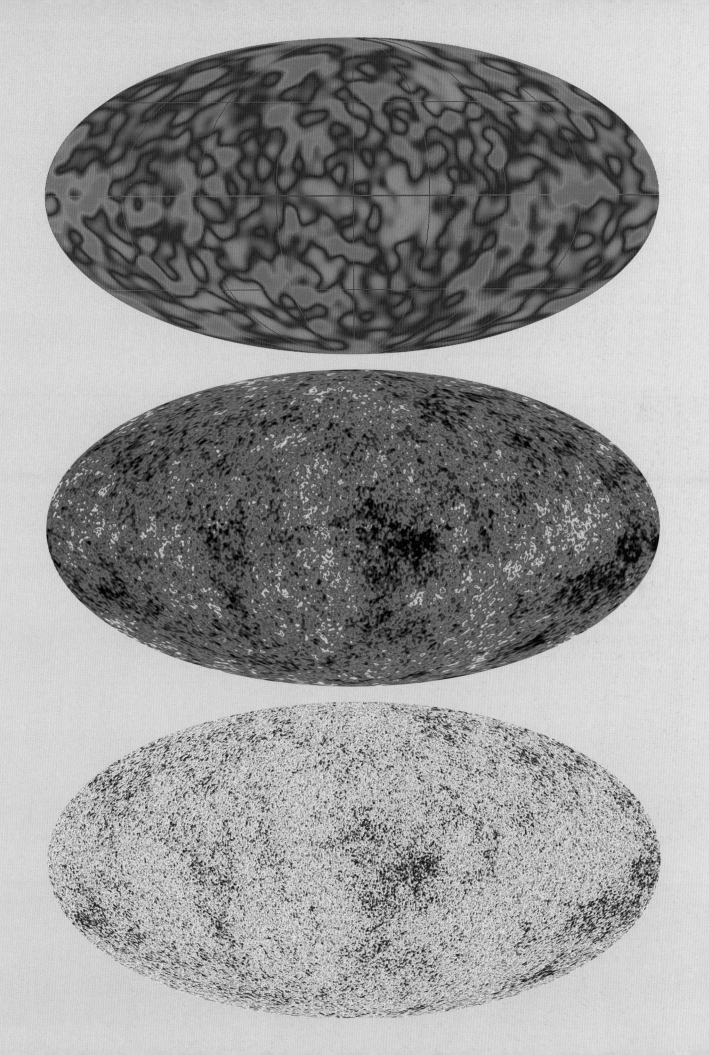

RIPPLES IN SPACE

Laser Interferometer Gravitational Wave Observatory and Evolved Laser Interferometer Space Antenna
Gravitational Waves

- In February 2016, on the centenary of their prediction, scientists announced the first ever detection of gravitational waves. In his revolutionary new theory of gravity, the general theory of relativity, Albert Einstein argued that when masses accelerate they will cause ripples in the very fabric of space. These ripples should spread out at the same speed as light travels, and at any given location space should expand and shrink as the waves pass by.

Since the 1960s, scientists have been looking for evidence of these gravitational waves. The challenge has been that the distortions in space are predicted to be truly tiny, requiring incredibly precise instruments to detect them. After several decades of planning and seeking funding, construction of the Laser Interferometer Gravitational wave Observatory (LIGO) began in the mid-1990s. Completed in 2002, LIGO consists of two gravitational wave detectors, both in the United States of America. One is located in Washington state, the other in Louisiana. Each observatory consists of two 4-kilometre long arms at right angles to each other. Light from a laser is split and passes along each arm; the light is reflected at the end of each arm to travel back to the origin. Any changes in the relative length of each arm can be detected by a change in the interference pattern of the light when it is recombined.

The instrument is so sensitive that it can detect relative changes in the length of each arm to less than a thousandth the width of an atomic nucleus. Any local vibrations, even the passing of vehicles on nearby roads, can cause changes. This is the reason for having two detectors separated by more than 3,000 kilometres; any real gravitational waves travelling at the speed of light will trigger the two detectors at different times, up to 10 milliseconds depending on the direction of the waves. The first such detection was announced in February 2016. LIGO had detected the passing of gravitational waves on 14 September 2015, and analysis showed that it was due to the merger of two black holes about 1.3 billion light-years from Earth.

Detecting gravitational waves provides an entirely new way of observing the Universe. The earliest light we can see is from the cosmic microwave background (CMB), as before this the Universe was opaque to any kind of electromagnetic radiation. The CMB originates from when the Universe was about 400,000 years old. Although we can infer things about the Universe before this time by their effects on details of the CMB, such as the size of the temperature variations and the polarization in the CMB, we have no way of directly observing the Universe earlier than this using any kind of electromagnetic radiation.

Gravitational waves, on the other hand, can travel to us from the very earliest fractions of a second after the Big Bang. They are unimpeded by the opacity which affects electromagnetic radiation, and so in theory we can use them to directly observe the Universe back to just after its creation. The challenge is how incredibly small the distortions in space due to primordial gravitational waves would be. We need an observatory thousands of times more sensitive than LIGO to detect them.

The solution proposed by the European Space Agency (ESA) is the Evolved Laser

Interferometer Space Antenna (eLISA). This ambitious project will put a constellation of three spacecraft into space, to form an equilateral triangle. The length of each arm will be one million kilometres; the array will fly around the Sun in an Earth-like orbit and the distance between each spacecraft will be precisely monitored to enable any passing gravitational waves to be detected. ESA has already launched the LISA Pathfinder mission in December 2015, the aim of this project is not to search for gravitational waves but to test some of the new technologies required for eLISA. With a proposed launch in 2034, eLISA is still several decades away, but should it come to fruition it will represent as exciting an advance in our abilities to study the Universe as the first telescopes gave us more than four hundred years ago.

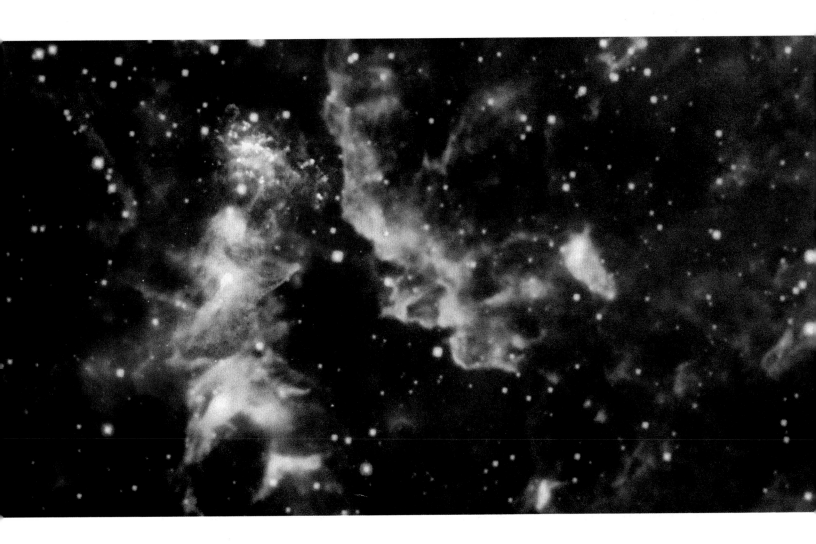

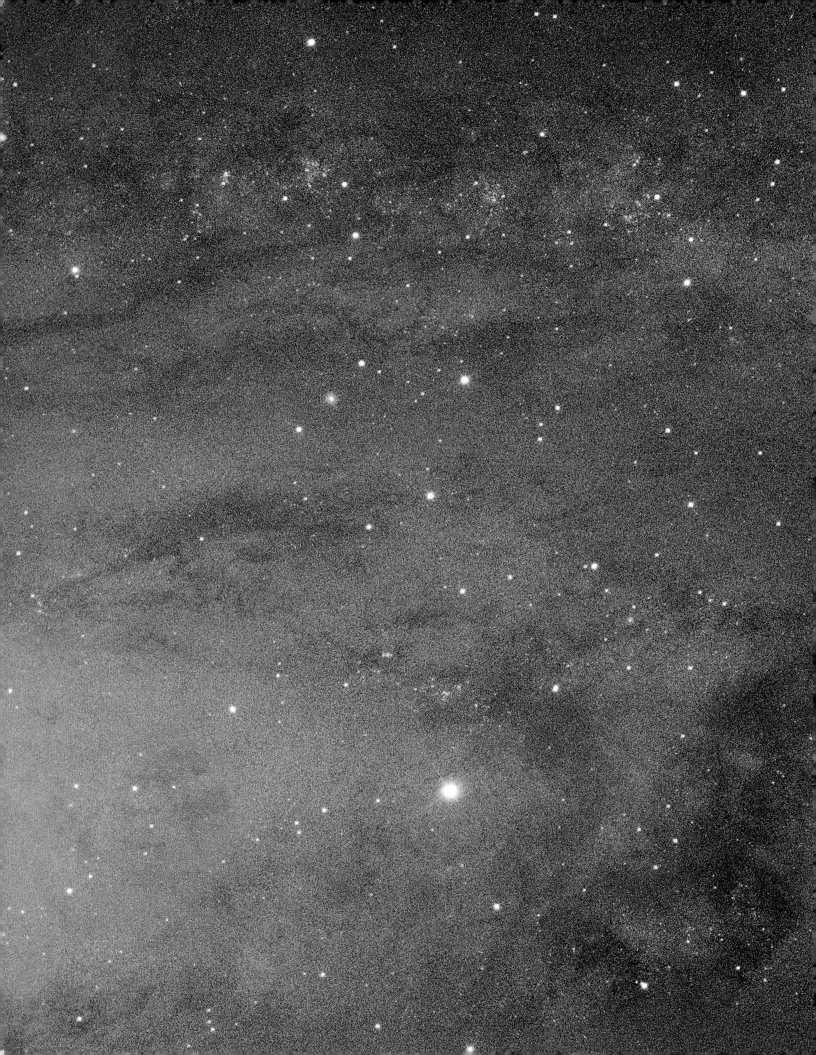

GLOSSARY

Absolute zero This is the coldest temperature theoretically possible. As temperature is a measurement of the motion of atoms, at absolute zero (0 Kelvin) all atomic motion would stop. In practice this is not possible, but in laboratories we have come within a few millionths of a degree of absolute zero.

Absorption spectrum This is when we see a continuous spectrum with a series of dark lines on it. The dark lines are produced by gases between us and the source of the continuous spectrum. These gases are absorbing radiation at particular wavelengths. For example, the gases in the atmosphere of the Sun produce an absorption spectrum, and by identifying the dark lines in the Sun's spectrum we can infer the temperature and pressure of the gases in its atmosphere. The element helium was first discovered in the absorption spectrum of the Sun before it had been detected on Earth, so it was named after "helios", the Greek name for the Sun. Measuring the wavelengths of the absorption lines in a spectrum allows us to determine whether the object producing the dark lines is moving towards or away from us, and at what speed it is moving.

Atom The ancient Greeks believed that matter could be broken down into indivisible units, which they called atoms. The modern idea of an atom is remarkably similar; it has a nucleus made up of positively charged particles called protons and neutral neutrons, which is surrounded by low-mass, negatively charged electrons. Electrons and protons have the same charge, although with opposite signs, and so a neutral atom must have an equal number of both. Carbon, for example, has six protons and six electrons. An atom of the lightest element – hydrogen – consists of a solitary proton and a single electron.

Atomic hydrogen The most common form of hydrogen in the Universe. In 1946 it was predicted that atomic hydrogen would radiate at 21-centimetres due to a transition in the ground state of the electron. This prediction led to the birth of radio astronomy in the late 1940s.

Asteroid Asteroids are rocky bodies which range from a few hundred kilometres in size to just a few tens of metres. Most asteroids lie in a belt between the orbits of Mars and Jupiter. Ceres, discovered in 1801, was the first asteroid to be found and is nearly 1,000 kilometres in diameter. In 2005 it was redesignated as a minor planet.

Aurorae Named after the Roman goddess of dawn Aurora, aurorae refers to the emission of light due to charged particles ionizing the gases in the atmosphere. On Earth we have the aurorae borealis (northern lights) and aurorae australis (southern lights). Aurorae have also been observed on other planets, including Jupiter.

Baryonic matter Protons and neutrons are examples of baryons. Astronomers use the term "baryonic matter" to distinguish the ordinary material in the Universe from the mysterious dark matter.

Big Bang theory First proposed by Georges Lemaître in the 1930s, the Big Bang theory argues that the Universe began at a particular moment in time, when the Universe and everything in it was created. The 1965 discovery of the cosmic background radiation led to its wide acceptance, and the competing Steady State theory has now been largely abandoned.

Billion Throughout this book we have used the standard scientific definition of billion, which is a thousand million (1,000,000,000 or 10^9). The older, English definition, according to which a billion equalled a million million, is now almost obsolete.

Blackbody An idealised emitter and absorber of radiation, to which a star approximates quite well. Any hot body emits electromagnetic radiation, and the spectrum of this emission for a blackbody is entirely determined by its temperature. A plot of energy against wavelength (or colour) for a blackbody produces a smooth "hump-backed" shape with a maximum intensity that moves to a shorter wavelength as the temperature increases. If a metal object such as a poker is heated, it appears first red, then orange, then yellow and eventually white hot. Similarly, the hottest stars have a blueish-white colour. The temperature of a blackbody can be precisely determined from the wavelength of maximum intensity of its spectrum.

Black hole A black hole is a body that has a strong enough gravitational field to prevent even light – the fastest thing in the Universe – from escaping. They were considered for many years to be theoretical curiosities, but there is now very strong evidence that black holes exist. In addition to black holes formed from the collapse of massive stars, it seems that most galaxies have a supermassive black hole (as massive as several million suns) at their centres. The origin of supermassive black holes remains a mystery.

Cepheid variable This is a particular type of star whose brightness varies, named after its archetype Delta Cepheus. It was realized in the early twentieth century that cepheid variables showed a relationship between their intrinsic brightness and their period of variability; brighter cepheids took more time to vary their brightness than fainter ones. The most famous cepheid variable star is Polaris, the North Star.

Charge Electric charge is a property shared by particles such as protons and electrons. Positive and negative electric charges attract each other, and it is this force that keeps the negative electrons bound to the positive nucleus in a neutral atom.

Comet An icy body, best described as a "dirty snowball". Comets are believed to come from two main sources. Short-period comets come from the Kuiper belt, which lies just beyond the orbit of Pluto. Longer period comets come from the Oort Cloud, which is a spherical distribution of material lying thousands of times further away than the Kuiper belt.

Constellation A group of stars that appear near each other in the sky, forming a recognizable pattern. The stars in a constellation have no physical link to each other and may be many thousands of light years apart. Although ancient cartographers chose to create their own constellations, in 1930 the International Astronomical Union selected 88 for their official list. The largest constellation is Hydra, the Sea Serpent, and the smallest is Crux, the Southern Cross. Although the constellations appear constant on the timescales of our human lifetimes, the stars are slowly moving and over time their familiar patterns will disappear.

Corona The thin outer atmosphere of the Sun, usually only visible during total eclipses. It emits mainly in X-rays, due to its very high temperature. How it gets to such a high temperature is currently a mystery.

Cosmic microwave background Discovered in the 1960s, the cosmic microwave background (CMB) originates from a time when the Universe had cooled enough for matter and radiation to decouple from each other. Since it was emitted, the Universe has expanded by a factor of about one thousand, stretching this visible wavelength radiation into the microwave part of the spectrum. The CMB provides one of the strongest pieces of evidence for the Big Bang theory.

Crater Formed by the impact of an object on the surface of a planet or moon. The number of craters can often be used as an indicator of the age of the surface of a planet or moon; absence of craters suggests that the planet or moon is geologically active.

Cryovolcano This is a volcano where the erupting material are volatiles such as water, ammonia or methane, rather than molten rock. Cryovolcanoes are found on several Solar System bodies including Neptune's moon Triton and Saturn's moon Enceladus.

Dark matter Over the last 50 years astronomers have come to realize that most of the matter in the Universe is composed not of ordinary atoms and molecules, but is in the form of some exotic material which we call dark matter. In fact, more than 80 per cent of the mass in the Universe is made up of dark matter, which interacts with "normal" (or baryonic) matter almost exclusively through gravity.

Dark nebula This is a nebula which is so dense that it blocks the light from behind it, showing up as a patch of darkness surrounded by light. One of the best known examples of a dark nebula is the Horeshead Nebula in the Orion constellation.

Doppler effect The Doppler effect is most familiar from the change of pitch heard in a siren as an ambulance drives past. The waves emitted from an approaching source are compressed, and hence appear to have a higher pitch than those from a stationary source. Conversely, waves emitted from a receding source are stretched and hence appear to have a much lower pitch. The greater the relative speed between the source and the observer, the greater the shift in wavelength. The same effect applies to light; the stretching of the light means that light from a receding source will appear reddened (called redshift), and that from an approaching source will appear shifted toward the blue end of the spectrum.

Dust Interstellar dust are tiny particles of mainly carbon or silicon which are formed by stars as the stars come towards the end of their lives. Dust absorbs and scatters starlight, and so hides many areas from our view at visible-light wavelengths. By looking in the infrared, we can often peer through this obscuring dust and see e.g. sites of active star formation.

Dwarf galaxy The smallest type of galaxy, dwarf galaxies are often hard to find due to their small size and low surface brightness. Several of the dwarf galaxies which orbit our own Milky Way Galaxy have been discovered only in the last few decades.

Dwarf planet After the discovery of several Kuiper belt objects, Pluto was reclassified as a dwarf planet in 2005. Ceres, the first asteroid to be discovered, has also been reclassified as a dwarf planet, as have Sedna and Eris, both Kuiper belt objects discovered since 2000.

Eclipse About twice a year, during a New Moon, the shadow of the Moon can fall on the Earth, creating an eclipse. When the alignment is just correct we have a total Solar eclipse; in the path of totality the Sun's disk is completely blocked by the Moon's disk, turning day into night and revealing the Sun's corona. During the Full Moon two weeks before or two weeks after a Solar eclipse, the Moon will move into the shadow of the Earth; we call this a Lunar Eclipse.

Electromagnetic radiation Visible light is just one part of the spectrum that stretches from ultra-high energy gamma rays and X-rays through the ultraviolet, then through the visible into the infrared, microwaves and then finally the radio wavelengths. Electromagnetic radiation in all of these forms is composed of an electric and a magnetic component at right angles to each other, which move at the speed of light.

Electron A low-mass particle (less than one thousandth [0.001] the mass of a proton) with unit negative electric charge. Unlike protons and neutrons, electrons are not made up of quarks and appear to be truly "fundamental" particles that cannot be broken into smaller parts.

Emission nebula A type of nebula whose spectrum shows a series of bright, narrow lines. These lines can be used to identify the composition of the nebula. The best known example of an emission nebula is the Orion Nebula (Messier 42).

Energy The law of conservation of energy (otherwise known as the first law of thermodynamics) is one of the most fundamental of all physical laws. It states that energy is neither created nor destroyed, but can only be converted from one form to another. The famous equation $E=mc^2$ simply states that mass can be converted into energy or, equivalently, that mass is simply another form of energy. The nuclear reactions at the centres of stars convert mass into radiation and thermal energy.

Equator The imaginary circle drawn on a sphere so as to be an equal distance from both poles. We are relatively familiar with the Earth's equator, and the projection of this line on to the sky defines the celestial equator. It is useful as a point of reference for our coordinate systems, but its position on the sky has no physical significance.

False colour Combining an image taken through red, green and blue filters creates a natural colour image. However, sometimes we assign different colours to, for example, radio waves, infrared or X-rays to help illustrate the emission from these different components. Such images are referred to as false colour images, as their colours are used merely to represent or highlight different components.

Frequency How often a wave oscillates every second. Light travels as a wave, and we can specify either its wavelength or its frequency. The speed of a wave is just its frequency multiplied by its wavelength, so they are inversely related. A high frequency means a short wavelength, a low frequency means a long wavelength.

Galactic halo This term refers to a sphere surrounding a galaxy, extending far beyond the visible component of stars. Studies of the motions of stars in spiral galaxies suggest that most of the mass of galaxies lies in the galactic halo, but rather than being in the form of baryonic matter, most of this mass is in the form of dark matter.

Galaxy From the Greek word for "milk", the term galaxy was first applied to the Milky Way, which was seen as a faint strip of light running through the sky. Once it became clear that our own Galaxy was only one of many billions, the term was applied to mean any large group of stars and other material that exists as an independent system, held together by its own gravity. The two major classes of galaxies are the ellipticals and the spirals. The ellipticals are large, spherical systems of old stars with relatively little gas remaining to be converted into stars. By contrast, a spiral is characterised by a disk containing spiral arms that mark ongoing star-forming activity surrounding a central, older bulge. It was believed for many years that elliptical systems formed from the collision of two spirals, but the process appears to be more complicated than that. Lenticular galaxies lie between these two main types; they are galaxies with a prominent bulge, but also a disk, although the spiral arms tend to be less prominent or absent compared to a spiral galaxy.

Gamma rays The most energetic part of the electromagnetic spectrum. Gamma rays are produced by intense processes such as radioactive decay and the collapse of neutron stars into black holes.

Gas giants The planets Jupiter, Saturn, Uranus and Neptune are referred to as gas giants, as they are mainly composed of the gases hydrogen and helium. At their centres they may contain a rocky core.

Geostationary orbit The closer a satellite is to the Earth's surface, the less time it takes to orbit the Earth. The International Space Station is about 500 kilometres from the Earth's surface, and takes about 90 minutes to orbit the Earth. A satellite which takes 24 hours to orbit the Earth is in a geostationary orbit; it will appear in the same place in the sky as seen from a particular location on Earth. This is particularly useful for communication satellites. The height of the geostationary orbit is about 35,000 kilometres above the Earth's surface.

Globular cluster A collection of hundreds of thousands of stars, globular clusters are found in the halo of our Milky Way Galaxy and in the halos of other galaxies. They lack any young stars, and are believed to be amongst the first objects to have existed in the formation of galaxies.

Gravitational lens Albert Einstein's general theory of relativity predicted that light would be bent by gravity. First seen in the 1980s, gravitational lenses occur when a foreground galaxy or cluster of galaxies distort and magnify a background galaxy. Gravitational lenses allow us to measure the distribution of matter in galaxy clusters, and provide one of the strongest pieces of evidence for the existence of dark matter.

Gravitational waves Predicted by Albert Einstein in his general theory of relativity, gravitational waves are distortions in space produced when massive bodies accelerate. The first ever direct confirmation of gravitational waves was announced in February 2016, when such waves were detected from the merger of two black holes.

Gravity Although gravity is intrinsically the weakest of the fundamental forces, it is the only one of the four forces that acts on astronomical scales. Of the others, the strong and weak nuclear forces act only within atomic nuclei, and the electromagnetic force from positive and negative charges cancels out. The gravitational attraction between two objects is proportional to their masses and inversely proportional to the square of their separation. In other words, two masses moved so that the distance between them is halved will attract each other four times as strongly. The first systematic theory of gravity was due to Sir Isaac Newton, whose theories were expanded by Albert Einstein in his general theory of relativity.

Gravity assist In order to help spacecraft gain speed, scientists often use a gravity assist. To do this, the spacecraft's trajectory sends it near to a planet; the planet's force of gravity transfers some of the planet's momentum to the spacecraft, which in turn speeds up the spacecraft. Because of the limited amount of fuel which a spacecraft can take with it when launched, gravity assists form a vital function in allowing us to send spacecraft to other parts of the Solar System.

Heat The scientific definition of temperature is rather different from the everyday one. The higher the temperature of a gas, the faster the atoms that compose it are moving. By contrast, "heat" is usually used to mean the quantity of thermal energy present. For example, a firework sparkler is at a much higher temperature than a red-hot poker, but because there is much less mass in a firework sparkler than a poker there is more heat in the poker – which is why one can hold a sparkler but would be reluctant to hold a glowing poker.

HII region HII is the astronomical nomenclature for singly ionized hydrogen. HII regions are often associated with sites of active star formation; ultraviolet light from the hot young stars ionizes the surrounding neutral hydrogen gas. When the electrons try to recombine with the protons they cascade down energy levels in the hydrogen atoms, producing well-known spectral lines including hydrogen-alpha (H-alpha).

Hubble constant The term given to the rate at which the Universe is expanding. It has been measured to have a current value of 72 kilometres per second per Megaparsec.

In the late 1990s it was found that the Hubble constant is greater now than when the Universe was about half of its current age, meaning that its expansion is speeding up. This was a big surprise, and the component responsible has been termed "dark energy", although the nature of dark energy remains a mystery.

Infrared-light Discovered in 1800 by accident by William Herschel, the infrared lies between the visible and the microwave part of the spectrum. It was the first part of the spectrum outside of visible light to be discovered. All warm objects radiate in the infrared part of the spectrum, including our own bodies. Longer wavelength infrared radiation is often from warm and hot dust grains, but shorter wavelength infrared radiation comes from cool stars, and this radiation is absorbed far less by dust than visible-light is.

Interstellar gas The space between stars is not empty, but contains gas and dust. Most of the interstellar gas is hydrogen, but other elements such as helium, oxygen, nitrogen, carbon and calcium are also observed.

Intracluster gas The extremely hot gas found between galaxies in clusters of galaxies. Due to its very high temperature, intracluster gas emits at X-rays, and is often responsible for most of the baryonic mass in a cluster.

Ionization Energetic photons can knock electrons away from their atomic nuclei, which are then said to be ionized. HII regions are areas where hydrogen gas has been ionized; but we also see radiation from many other ionized elements such as oxygen, calcium, iron etc.

Kelvin In science we use the Kelvin temperature scale, where absolute zero (-273 Celsius) is defined as 0 Kelvin. The size of a Kelvin is the same size as a degree Celsius, so 0 Celsius is +273 Kelvin.

Kuiper belt A reservoir of material just beyond the orbit of Pluto, which is the source of short-period comets. It was first proposed by Gerard Kuiper in the 1950s, and astronomers began to discover several in the 1990s. It is now accepted that Pluto is a Kuiper belt object, and in 2005 it was reclassified as a dwarf planet.

Lagrangian Point There are five Lagrangian points for the Earth-Sun system, named after Italian mathematician Joseph-Louis Lagrange. Of particular interest are the L1 and L2 points, which lie on a straight line with the Earth-Sun line. L1 lies closer to the Sun than the Earth, L2 lies further away. At both L1 and L2 an object will take the same time to orbit the Sun as the Earth takes, making them particularly useful locations for placing astronomical satellites.

Light-year The distance travelled by light in a year when passing through a vacuum, equivalent to 9.5×10^{15} metres or nearly six thousand billion miles. The Sun is eight light-minutes away – the light we see left the Sun eight minutes before – and the nearest other star is 4.2 light-years away. The Sun lies 26,000 light-years from the centre of the Milky Way Galaxy, which is itself 100,000 light-years across. Bodies located 13 billion light-years away are seen as they appeared just after the Big Bang.

Local Group The group of galaxies of which our Milky Way Galaxy is a member. Its other large member is the Andromeda Galaxy, but it also contains a number of smaller members such as the Large and Small Magellanic Clouds, Messier 33 and dwarf galaxies such as Sextans A and NGC 185.

Luminosity The luminosity of a light source is related to the rate of emission of light. In other words, the luminosity of a star depends on its intrinsic rather than apparent brightness. The Sun appears much brighter than the other stars in the sky because it is close to us, even though many stars are much more luminous than our own rather ordinary star.

Magnitude The traditional measure of brightness for astronomical objects. The scale is rather confusing; the lower the number, the brighter the source appears. By definition, the bright star Vega has a magnitude of 0.0 and a difference of five magnitudes corresponds to a difference of 100 times in brightness. Vega is therefore 100 times brighter than a star with a magnitude of 5. In dark skies, the naked eye can see to about a magnitude of 6. These are apparent magnitudes, but it is also common to refer to absolute magnitudes. These reflect the luminosity of the source and are defined as the apparent magnitude the source would have at the standard distance of 10 parsecs.

Mass There are two scientific definitions of mass. The first is the property of a body to resist acceleration; it takes more effort to push a car than a football. The second is the property of a body that defines the strength of its gravitational attraction; objects with more mass have a stronger gravitational pull. The two turn out to be equivalent so that the same definition of mass can be used for both. A common mistake is to confuse mass with weight. Weight is the force exerted on an object by gravity. When Neil Armstrong stepped out onto the Moon's surface, his mass did not change but his weight certainly did.

Messier Catalogue Many of the objects in this book are found in the Messier Catalogue, compiled by comet hunter Charles Messier in the 18th Century. Messier wanted to catalogue fuzzy objects which were not comets, so that he and other comet hunters would not mistake them for comets. For example, Messier 1 is the Crab Nebula, Messier 42 is the Orion Nebula, Messier 31 is the Andromeda Galaxy.

Microwaves These lie between radio waves and infrared light in the electromagnetic spectrum. Microwaves are very effectively absorbed by water, so to best observe the cosmic microwave background we need to send satellites into space.

Milky Way A luminous band of faint stars that crosses the sky, containing many nebulae and dust clouds in addition to stars. It is the projection of the disk of our own Galaxy, which is also known as the Milky Way, onto the celestial sphere.

Millimetre waves Lying at the short end of the radio spectrum, between microwaves and radio waves, are millimetre waves. Many astrophysically important molecules including carbon monoxide radiate in this part of the spectrum.

Molecular cloud Molecular clouds are composed primarily of molecular hydrogen. They are the main sites of star formation as their centres are well shielded from stellar radiation, providing a cool environment where gas and dust can collapse to form new stars.

Moon A moon is an object that orbits a planet. Our Earth has one moon, which we call the Moon. The only planets in our Solar System which do not have moons are Mercury and Venus.

Nebula From the Latin word for "mist", "fog" or "cloud", the term "nebula" is used in astronomy to refer to any visible mass of gas and dust. The most famous nearby example, the Orion Nebula, is a region in which stars are forming from condensing gas and dust. Many famous nebulae are now known to be galaxies; the best example is the Andromeda galaxy, previously known as the Andromeda Nebula.

Neutron star Neutrons are one of the two types of particles – the other being protons – that make up atomic nuclei. They weigh almost the same as protons, but carry no electric charge. Under the extreme conditions of a supernova explosion, protons and electrons can combine to form neutrons, resulting in a dense neutron star being produced from the dying star's core. The maximum mass for a neutron star is believed to be around eight solar masses; any larger than this and it will collapse into a black hole.

NGC Many objects in the sky have an NGC designation. The acronym stands for a catalogue of objects listed in the *New General Catalogue of Nebulae and Clusters of Stars*, complied in 1888 by John Louis Emil Dreyer. This catalogue was an update of the *General Catalogue of Nebulae and Clusters of Stars*, complied in 1864 by John Herschel, William Herschel's son. There are 7,840 objects in the NGC.

Nucleus The nucleus of an atom is made up of positively charged protons and neutral neutrons, and contains almost all of the mass of the atom. At the high temperatures and pressures in the centres of stars, electrons are too energetic to be captured by the positively charged nucleus; so it is atomic nuclei that combine in nuclear fusion to form heavier elements. The number of protons in the nucleus of an atom defines its type, so that hydrogen has a single proton, helium two, lithium three and so on.

Parsec A unit of distance equal to 3.26 light-years. A star one parsec away will exhibit a parallax shift of 1 arc-second as our Earth orbits the Sun and we view it from two opposite vantage points.

Photon The name given for a particle of electromagnetic radiation is a photon; this term was coined in the 1920s. A photon is an elementary particle, the quantum of all forms of electromagnetic radiation, including light. In the modern-day model, electromagnetic radiation travels as a wave but interacts with matter as a particle.

Planet Originally from the Greek word for a "wandering star", the modern definition of a planet is an object which orbits a star. Our Solar System contains eight known planets; Pluto was redesignated a minor planet in 2005. Since 1995 we have discovered planets around other stars, which we call extra-solar planets (or exoplanets). Since 1995, more than a thousand extra-solar planets have been discovered.

Proton A positively charged particle composed of three quarks; protons are one of the two components that make up atomic nuclei, the other being neutrons.

Pulsar A rapidly spinning neutron star (produced in a collapsing supernova) will produce radiation in a thin beam from near both poles. As the star rotates, so this beam sweeps, lighthouse-like, across the sky. If it happens to cross the Earth, we see a rapidly pulsing source. So regular are these pulses that the first detection was labelled "LGM-1", standing for Little Green Men-1! There is one known example of a double pulsar, and scientists are able to exploit the information from its pulses to provide stringent tests of the theory of general relativity.

Quasar The original definition of quasar, or quasi-stellar object, was a star-like source that appeared to be at a great distance. Decades of observations have revealed that it is in fact a galaxy harbouring an extremely massive black hole at its centre, which is in the process of consuming huge amounts of dust, gas and stars. This in-falling material radiates as it falls toward the central black hole, and this powerful source of radiation is responsible for our ability to see quasars from the most distant parts of the Universe. They were more common in the distant past, and it has recently been suggested that all galaxies may have experienced a "quasar-like" phase, relaxing to become "normal" galaxies only when the reservoir of material to feed the central black hole has been exhausted.

Radio waves The radiation with the lowest energy in the electromagnetic spectrum. The first radio waves to be discovered coming from space were in the 1930s, when Karl Jansky found radio emission from the centre of the Milky Way Galaxy. Neutral hydrogen emits very low energy radio waves, with a wavelength of 21 centimetres. This realization, made in 1946, led to a boom in radio astronomy from the 1950s.

Redshift The movement of spectral features toward the red end of the spectrum of a receding source, due to the Doppler effect. Because of the expansion of the Universe, distant objects exhibit a larger redshift than nearer ones. Redshift is related to the size of the Universe, so when we observe a galaxy with a redshift of z=1 we are seeing the Universe when it was about half of its current age. The most distant galaxy observed at the time of writing this has a measured redshift of z=11.1.

Reflection nebula These have a characteristic blue appearance. They are produced when dust lying to the side of a bright star scatters (or reflects) the starlight towards us. Blue light is scattered much more than red light, and so the reflection nebula usually appears more blue than the illuminating star.

Relativity Albert Einstein published two landmark theories which have become known as the special theory of relativity (or just "special relativity") and the general theory of relativity (or just "general relativity"). Special relativity, published in 1905, deals with objects travelling at a constant velocity, and showed that lengths contract and time dilates at speeds close to the speed of light. The famous equation $E=mc^2$ was developed by Einstein a few months later as a natural consequence of special relativity. General relativity was published in 1916, and deals with the case of acceleration. It argued that acceleration and gravity were equivalent, and reinterpreted gravity as being due to masses curving the fabric of space.

Sol This is the term used for the length of a day on Mars. Mars rotates once on its axis every 24 hours, 39 minutes and 35 seconds, slightly longer than an Earth day. Of all of the planets in the Solar System, the length of a day on Mars is most similar to the length of a day on Earth. Mars lies roughly 1.5 times further from the Sun than the Earth, and so its orbit around the Sun takes longer. It takes Mars 687 Earth-days to orbit the Sun.

Solar flare An eruption from the surface of the Sun, usually associated with sunspots. A particularly large solar flare is called a coronal mass ejection, and these can lead to increased auroral activity on Earth.

Spectrum Electromagnetic radiation passed through a prism (or a fine grating) will split into its component wavelengths, an effect most familiar from the sight of a rainbow in the sky. This is known as a spectrum, and the relative intensities of different wavelengths can encode a huge amount of information about the object that emitted the light. In particular, a series of dark or bright lines known as spectral lines acts as a fingerprint for each of the elements present in the source, allowing astronomers to identify the composition of even the most distant objects. Sir Isaac Newton coined the word spectrum from the Latin for "to see". There are three types of spectra. A continuous spectrum has emission at a range of wavelengths with no gaps in the brightness of the spectrum. A continuous spectrum is produced by a hot opaque solid, liquid or gas. An emission spectrum is when we see a dark background with a series of bright lines at discrete wavelengths. An emission spectrum is produced by a thin liquid or gas, and by identifying the lines in the emission spectrum it is possible to determine the elements and molecules which make up the liquid or gas. Finally, an absorption spectrum is produced by the same thin liquid or gas which produces an emission spectrum; we see the spectrum as an absorption spectrum when we view it against a continuous spectrum. If we look from a different direction where the source of the continuous spectrum is not in the line of sight we see an emission spectrum.

Steady State theory A now discredited rival to the Big Bang theory, which held that the Universe had always existed and was in a constant state of continued expansion with matter being continually created.

Stellar wind Our Sun gives off a stream of charged particles which we call the Solar wind. Many hot stars are found to emit copious quantities of charged particles; we call these stellar winds. Along with ultraviolet light, stellar winds are responsible for ionizing interstellar gas and eroding the surfaces of molecular clouds.

Supercluster This is the term used for clusters of clusters of galaxies. It was the astronomer Vera Rubin who, in the 1950s, first realized that clusters of galaxies themselves formed structures of even larger sizes; we now call these superclusters.

Supernova In the 1940s Fritz Zwicky and Walter Baade coined the term supernova to refer to a massive star ending its life in a dramatic explosion. Our Sun is not massive enough to become a supernova. When a supernova explodes it can become brighter than all the other stars combined in its host galaxy. It is in supernova explosions that all the elements heavier than iron are formed, and this processed material is thrown into space and will be used in a future generation of stars.

Supernova remnant The material left after a supernova explosion. The first object to be recognized as a supernova remnant was the Crab Nebula; we now know that it is the remnant of a star which was seen to explode in 1054. Supernovae remnants contain material rich in elements which have been processed within the progenitor star and its explosion. This enriched material is reprocessed into a future generation of stars and planets.

Surface brightness A measurement of the amount of light emitted per unit area for a galaxy. Galaxies with a high surface brightness can often be mistaken for stars, and galaxies with a low surface brightness are hard to find as their light is lost against the glow of the sky.

Telescope Galileo Galilei is often incorrectly attributed with the invention of the telescope. In fact, the telescope was invented in the Netherlands and Galileo heard about it from a friend, whereupon he quickly built his own based on his friend's description. There are two basic types of telescopes; a refracting telescope uses a lens to gather the light, a reflecting telescope uses a mirror. The largest refracting telescope in existence is the 40-inch (1-metre) refractor at Yerkes Observatory, part of the University of Chicago. The largest reflecting telescopes currently have mirrors of as large as 10 metres, often using segmented mirrors. There are plans to build telescopes with segmented mirrors as large as 30 metres in the 2020s. It is the light gathering power of a telescope that is important; this depends on the size of its lens or mirror. Magnification, often used to advertise cheap telescopes, is of little importance in astronomy.

Terrestrial planets These are the four planets of the inner Solar System; Mercury, Venus, Earth and Mars. They are very different in composition from the gas giants found in the outer Solar System.

Tides This term has a more precise meaning in astronomy than the familiar rise and fall of the sea on Earth due to our Moon. An extended body can feel tidal forces when different parts of it experience a different force of gravity. The extreme volcanic activity on Io, Jupiter's moon, is due to Jupiter's gravity creating tidal forces in Io as it orbits Jupiter in an elliptical path; these tidal forces stretch and squeeze Io and generate a huge amount of internal heat in the moon. When galaxies gravitationally interact, the material in the galaxies is often disrupted by tidal forces; the tidal tails of the Antennae Galaxies are a good example.

Ultraviolet light This light lies just beyond the blue end of the visible part of the spectrum. Discovered in 1801, the year after the discovery of infrared-light, ultraviolet (UV) light is emitted by hot young stars. Such light is energetic enough to ionize neutral hydrogen, leading to the HII regions which we often see near sites of recent star formation.

Visible light A tiny part of the electromagnetic spectrum to which our eyes are sensitive. It ranges from blue light at about 0.3 microns to red light at about 0.7 microns. Until the 1950s, all of astronomy was done using only visible-light.

Wavelength The distance between two crests of a wave. The wavelength of red visible light is 4.0×10^{-7} metres (0.4 microns), while radio waves can have wavelengths of many kilometres.

X-rays After gamma rays, these are the next most energetic form of electromagnetic radiation. In space they can be produced by extremely energetic processes such as the very hot gas in between galaxies in clusters (the intracluster medium).

INDEX

CREDITS

Every effort has been made to acknowledge correctly and contact the source and/or copyright holder of each picture. Any unintentional errors or omissions will be corrected in future editions of the book. Please contact Canopus Publishing.

8 NASA/JPL
10–11 INT Photometric H-Alpha Survey (IPHAS), Nick Wright (University of Hertfordshire, SAO)
12 NASA/JPL
14–15 ESO/Y. Beletski
16 Yohkoh/ISAS/Lockheed-Martin/NAOJ/ University of Tokyo/ NASA
17 STEREO/ESA/NASA/SOHO
19 t. SOHO b. NASA/ISS
20–21 NASA/Johns Hopkins University Applied Physics Laboratory/ Carnegie Institution of Washington
22 NASA/JPL/USGS
23 Courtesy of the NAIC – Arecibo Observatory, a facility of the NSF
24–25 NASA
26 NASA/JPL-Caltech/MSSS
27 NASA/JPL/University of Arizona
28–29 NASA/JPL/University of Arizona
30 t. NASA/JPL/Cornell b. NASA/JPL-Caltech/University of Arizona
31 NASA/JPL-Caltech/University of Arizona
32 NASA, ESA, and M. Kornmesser. Science Credit: NASA, ESA, L. Roth (Southwest Research Institute and University of Cologne, Germany), J. Saur (University of Cologne, Germany), K. Retherford (Southwest Research Institute), D. Strobel and P. Feldman (Johns Hopkins University), M. McGrath (Marshall Space Flight Center), and F. Nimmo (University of California, Santa Cruz)
33 t. NASA b. NIMS, Galileo Mission, JPL, NASA
34–35 NASA/ESA/John Clarke (University of Michigan)
36 NASA/JPL
37 NASA/Johns Hopkins University Applied Physics Laboratory/ Southwest Research Institute
38 t. NASA/HST b. NASA/JPL-Caltech/SETI Institute
39 NASA/JPL-Caltech/SETI Institute
40 NASA/JPL-Caltech/Space Science Institute and NASA/Johns Hopkins University Applied Physics Laboratory/Carnegie Institution of Washington
41 ESA/NASA/JPL/University of Arizona, image processing and panorama by René Pascal
42 t. NASA/JPL-Caltech/Space Science Institute b. NASA/Johns Hopkins University Applied Physics Laboratory/Carnegie Institution of Washington
43 NASA
44–45 NASA/JPA/Caltech/SSI
46–47 NASA/JPL/SSI/Gordan Ugarkovic
48–49 b. Voyager Project, JPL/NASA t. Voyager Project, JPL/NASA
50–51 NASA/JPL/USGS
52–55 NASA/JHUAPL/SWRI
56–57 Dan Burbank/NASA/ISS
58 ESA/Rosetta/MPS for OSIRIS Team MPS/UPD/LAM/IAA/SSO/ INTA/UPM/DASP/IDA
59 t. ESA/Rosetta/Philae/CIVA b. ESA/J Huart
60 t. ESA/Rosetta/Philae/CIVA b. CIVA/PHILAE/ROSETTA/ESA
61 ESA/Rosetta/NAVCAM
62 Lucent Technologies' Bell Laboratories, courtesy AIP Emilio Segre Archives
64 NASA
65 2MASS, Umass, IPAC/Caltech, NSF, NASA
66–67 NASA/JPL-Caltech/E Churchwell (University of Wisconsin), GLIMPSE team, S Carrey (SSC Caltech), MIPSGAL team

68–69 Atlas Image mosaic obtained as part of the Two Micron All Sky Survey (2MASS), a joint project of the University of Massachusetts and the Infrared Processing and Analysis Center/ California Institute of Technology, funded by NASA and NSF
68 m. Reid Wiseman/NASA b. NASA/GSFC/COBE
70 ESO/J. Emerson/VISTA. Acknowledgment: Cambridge Astronomical Survey Unit
71 Akira Fujii
72–73 NASA/HST
74–75 ESO/M.-R. Cioni/VISTA Magellanic Cloud survey. Acknowledgment: Cambridge Astronomical Survey Unit
76–77 ESO/Igor Chekalin
78 NASA/JPL/Caltech
79 NASA/ESA and the Hubble Heritage Team (StScI/AURA)
80–81 ESA/Herschel/NASA/HST
81 t. NASA/HST
82–83 NASA/CXC/SAO (X-ray), Paul Scowen and Jeff Hester (Arizona State University) and the Mt. Palomar Observatories (optical), 2MASS/UMass/IPAC-Caltech/NASA/NSF (infrared), and NRAO/AUI/NSF (radio)
83 t. NASA/ESA and The Hubble Heritage Team STScI/AURA)
84 NASA/ESA/AURA/Caltech
85 NASA/JPL-Caltech/J. Stauffer (SSC/Caltech)
86 ESO
87 ESO
88 NASA/ESA Hubble Heritage Team/StSci/AURA
89 NASA/ESA/STSci
91 NASA/JPL-Caltech/2MASS
92 NASA/JPL-Caltech/ESA/CXC/STScI
93 NASA/UMass/D.Wang et al
94–95 ESO
96 AIP
98–99 NASA/HST
98 t. B Wakker (University of Wisconsin Madison) et al., NASA
100–101 NASA/JPL-Caltech/K. Gordon (University of Arizona) & GALEX Science Team
102 X-ray: NASA/CXC/University of Potsdam/L. Oskinova et al; Optical: NASA/STScI; Infrared: NASA/JPL-Caltech
103 David Malin, AAT
104–105 ESO
106–107 NASA/HST
108–109 AURA/STSci/NASA/ESA
110 AURA/STSci/NASA/ESA
111 David Malin/AAT
112–113 NASA, ESA and M Livio (STScI)
114 Subaru Telescope/NOAJ
115 Dr. Hideaki Fujiwara – Subaru Telescope, NAOJ
116 X-ray: NASA/CXC/CfA/R. Tuellmann et al.; Optical: NASA/ AURA/STScI
117 Palomar Sky Survey
118 NASA
120 NASA/HST
121 Hubble data: NASA, ESA, and A. Zezas (Harvard-Smithsonian Center for Astrophysics); GALEX data: NASA, JPL-Caltech, GALEX Team, J. Huchra et al. (Harvard-Smithsonian Center for Astrophysics); Spitzer data: NASA/JPL/Caltech/Harvard-Smithsonian Center for Astrophysics
122–123 NASA/HST
124 NASA/JPL-Caltech/STScI/CXC/UofA/ESA/AURA/JHU
125 NASA/HST
126–127 Spitzer (infrared): NASA/JPL-Caltech/R. Kennicutt (University of Arizona), and the SINGS Team. Hubble (visible): NASA/Hubble

128–129 NASA/ESA and The Hubble Heritage Team (STScI/AURA)
130 NASA/JPL-Caltech/R. Kennicutt (University of Arizona/Institute of Astronomy, University of Cambridge) and the SINGS Team
131 Subaru Telescope/National Astronomical Observatory of Japan
132–133 : ESO/WFI (Optical); MPIfR/ESO/APEX/A.Weiss et al. (Submillimetre); NASA/CXC/CfA/R.Kraft et al. (X-ray)
136 ALMA (ESO/NAOJ/NRAO). Visible light image: the NASA/ESA /HST
137 Brad Whitmore (STScI), NASA/ESA
139 NASA/ESA/CXC/JPL/Caltech/STScI
140 NASA, ESA, and the Digitized Sky Survey
Acknowledgment: Z. Levay (STScI) and D. De Martin (ESA/HST)
141 Chris Mihos (Case Western Reserve University)/ESO
142–143 NASA/ESA/Hubble Herritage Team (STScI)
144 t. NASA/JPL/Caltech/SSC b. Block/Mount Adam Lemmon Sky Center/University of Arizona
145 NASA/Caltech/JPL
146–147 ESA/HST/ESO
148–149 ASA, ESA, and the Hubble SM4 ERO Team
150–151 NASA, ESA, the Hubble Heritage (STScI/AURA)–ESA/ Hubble Collaboration, and W. Keel (University of Alabama)
152 ALMA (ESO/NAOJ/NRAO)/NASA/ESA/F. Combes
153 NASA/HST
154–155 NASA/JPL/Caltech/GFSC/SDSS
156 X-ray: NASA/CXC/KIPAC/S.Allen et al.; Radio: NRAO/VLA/G. Taylor; Infrared: NASA/ESA/McMaster University/W.Harris
157 NOAO/AURA/NSF
159 t. ESA/HST/NASA b. ESA/HST
160–161 NASA/CXC/University of Missouri/M.Brodwin et al; Optical: NASA/STScI; Infrared: JPL/Caltech
161 b. NASA/HST
162 ESA/HST
163 NASA, ESA, E. Jullo (JPL/LAM), P. Natarajan (Yale) and J-P. Kneib (LAM)
164 X-ray: NASA/CXC/UCDavis/W.Dawson et al; Optical: NASA/ STScI/UCDavis/W.Dawson et al.
165 X-ray: NASA/CXC/CfA/ M.Markevitch et al.; Lensing Map: NASA/STScI; ESO WFI; Magellan/ UNIVERSITY OFArizona/ D.Clowe et al. Optical: NASA/STScI; Magellan/U.Arizona/D.Clowe et al.
166 ESA/HST/NASA
167 X-ray: NASA/CXC/UA/J. Irwin et al; Optical: NASA/STScI
170 NASA, ESA, H. Teplitz and M. Rafelski (IPAC/Caltech), A. Koekemoer (STScI), R. Windhorst (Arizona State University), and Z. Levay (STScI)
171 t. NRAO/AUI b. ESA/SPIRE/H-ATLAS/H.L. Gomez/
172 NASA/ESA
173 NASA / JPL-Caltech/A. KASHLINSKY (GSFC)
174–175 NASA, ESA and P. Oesch (Yale University)
174 b. NASA
176–177 NASA; ESM; G. Illingworth, D. Magee, and P. Oesch, University of California, Santa Cruz; R. Bouwens, Leiden University; and the HUDF09 Team)
178 t. ALMA (NRAO/ESO/NAOJ); B. Saxton NRAO/AUI/NSF; NASA/ESA HST, T. Hunter (NRAO) b. NASA/HST
179 ESO
181 COBE Project/DMR/NASA; Planck Collaboration/ESA; WMAP Science Team/NASA
183 NASA/CXC/SAO (X-Ray); NASA/JPL-Caltech (Infrared)
184–185 NASA, ESA, J. Dalcanton (University of Washington, USA), B. F. Williams (University of Washington, USA), L. C. Johnson (University of Washington, USA), the PHAT team, and R. Gendler.

THIS IS AN ANDRE DEUTSCH BOOK

Published in 2019 by André Deutsch
An imprint of the Carlton Publishing Group
20 Mortimer Street, London W1T 3JW

Text © Rhodri Evans 2016, 2019
Design © Carlton Books Limited 2016, 2019

Created for Carlton Books by Canopus Publishing Limited
www.canopusbooks.com

For Canopus Publishing Limited:
Director and Editor: Robin Rees
Additional Editorial: Tom Jackson
Design: Jamie Symonds

For Carlton Books Limited:
Editorial Director: Piers Murray-Hill
Editorial Manager: Alison Moss
Senior Art Editor: James Pople
Production: Sarah Kramer

A CIP catalogue for this book is available from the British Library.

ISBN 978-0-233-00579-9

Printed in Dubai

10 9 8 7 6 5 4 3 2 1